ECONOMICS AND TECHNOLOGICAL CHANGE

ECONOMICS AND TECHNOLOGICAL CHANGE

Rod Coombs
Paolo Saviotti
and
Vivien Walsh

Rowman & Littlefield
PUBLISHERS

ROWMAN & LITTLEFIELD

Published in the United States of America in 1987
by Rowman & Littlefield, Publishers
(a division of Littlefield, Adams & Company)
81 Adams Drive, Totowa, New Jersey 07512

Library of Congress Cataloging-in-Publication Data

Coombs, Rod.
  Economics and technological change.

  Bibliography: p.
  Includes index.
  1.Technological innovations—Economic aspects.
2. Industrial organization. 3. Technological innovations
—Government policy. 4. Technology and state—Citize
participation. I. Saviotti, Paolo. II. Walsh, Vivien.
III. Title.
HC79.T4C67   1987   338'.06   86–28000
ISBN 0–8476–7545–9
ISBN 0–8476–7546–7 (pbk.)

90 89 88 87
10 9 8 7 6 5 4 3 2 1

Printed in Hong Kong

Contents

Contents

Tables

Figures

Foreword

In recent years there have been a number of new books on economics of innovation, and while this is a welcome trend some of those have been a little repetitive. This one breaks new ground, and I believe it is the best and most original synthesis of the subject so far available.

Like all such books it is necessarily based on the research results obtained by a wide range of researchers in various countries. The authors have themselves all contributed substantially to this research literature. They have also taught courses in Manchester on the management of innovation, the economics of innovation and related topics to a variety of students. As a result of this experience, they have thought deeply about the problems in their field and their synthesis is a creative one, original both in detail and in the mode of presenting the entire subject.

In reading the book I could not help recalling the textbooks of economics which I was obliged to read as a student fifty years ago. They had scarcely a mention of research, development, innovation or diffusion of innovation. All the excitement and dynamism of real-world economic development was missing. To learn about what was actually happening in industrial organisations, in the emergence of new industries like electronics or in military research, we had to turn to biologists like Huxley and physicists like Bernal.

Now the picture is changing. Some modern economics textbooks do actually have a few sections or even a whole chapter on technical innovations. But the process still has a long way to go. Whether in the theory of the firm, or in the theory of international trade, the theory of consumer behaviour, or the theory of business cycles, obsolete dogmas which ignore technical innovation continue to be expounded and taught as though they were holy scriptures. This edifice of traditional economics has become a barrier to the understanding of the important processes of change in the modern world. Economic theory must end its neglect of both technical innovations and institutional change. This book is a major step forward in this direction.

PROFESSOR CHRISTOPHER FREEMAN

Science Policy Research Unit
University of Sussex

Preface

This book grew out of the involvement of the authors in the study of technological innovation from a variety of disciplinary standpoints. These standpoints are diverse as a result of the manifest range of institutions, economic and non-economic, with which innovation is connected. We found that the organisation of research work on innovation, and the organisation of teaching on the subject required us to have some 'view' on the ways in which these perspectives might be usefully integrated. This book certainly does not achieve that integration, but it does attempt to state the problems clearly, and to indicate the areas where advances have been made.

In a book of this nature, some comments on structure and purpose are appropriate. The book is divided into three parts. The first part deals primarily with the microeconomic and managerial dimensions of innovation within the institution of the firm. The purpose of this part is to situate innovation within the theoretical and descriptive accounts of firm behaviour which are current in the literature. The second part of the book examines the patterns in innovative behaviour which can be discerned at the level of industry, market and national economy. The purpose is therefore to explore the reciprocal relations between the institutions of the firm and the market in the determination of technical change. The third part of the book is concerned with the relationships between the process of technical change and the interests and actions of government, trade unions, pressure groups, and other institutions which do not reduce strictly to the firm or the market. These three types of institution: the firm, the market, and the 'civil' form an organising framework for the presentation of the analysis. However, as we stated above, our objective has been to explore the interactions as well as the distinctions between those attributes of technical change which can be referred to these three institutional contexts.

The material in the book is related to that taught to final year undergraduates and postgraduates in the Department of Management Sciences, UMIST, and in the Department of Science and Technology Policy, University of Manchester. The responses of those students over a period of some six years have played a valuable part in organising the material, for which we are grateful.

In writing the book we have struggled with the aim of satisfying the audiences of researchers, postgraduate and advanced undergraduate students, practitioners, and those with a general interest in public policy. Inevitably some of these audiences will find some sections either too cryptic or too long-winded. But we hope that all audiences will find substantial parts of the book pitched at or near an appropriate level.

Many colleagues have contributed to the book by stimulating us to examine particular issues, by criticising our ideas, and by writing some of the important work on which we attempt to build. In particular we would like to thank Christopher Freeman, Ken Green, Richard Whitley, Stanley Metcalfe, Michael Gibbons, Tony Cockerill, Trefor Jones, Alison Young, Roger Williams and Robin Roy, whilst absolving them of any responsibility for the way we have used their contributions. We also wish to thank Yvonne Aspinall for typing the book with great speed and accuracy.

ROD COOMBS
PAOLO SAVIOTTI
VIVIEN WALSH

Manchester

Acknowledgements

Grateful acknowledgement is made for permission to reproduce the following figures and tables: to the Econometric Society for figure 5.4 from Z. Grilicles, 'Hybrid Corn: An Exploration in The Economics of Technical Change', *Econometrica* 1957, p. 502; to Pergamon Press for figure 5.5 from I. Hendry, 'Three parameters approach to long range planning', *Long Range Planning* 1972; to Cambridge University Press for tables 7.1, 7.2, and 7.3 from D. Gordon, R. Edwards, M. Reich, *Segmented Work, Divided Workers* (1982); to C. Freeman, J. Clark and L. Soete for table 7.4 from their *Unemployment and Technical Innovation* Frances Pinter, 1982); and to Longmans Ltd, for table 4.1 from B. Twiss, *Managing Technological Innovation* (1980).

Part I
Technological Innovation and the Firm

1 Introduction

1.1 CHANGES IN THE PERCEPTION AND ANALYSIS OF TECHNOLOGICAL CHANGE

There is now a considerable and growing interest in research on technological change, particularly in some of the problems that this research has generated. Interest in technological change, however, has not been constant throughout history and perceptions of technological change have undergone very considerable variations in the course of time. These variations are perhaps related to the changing role of technological change in socioeconomic development. Changes in technology have always been an important component in the progress of human societies; as long ago as the invention of the wheel and discovery of fire, and more recently in the development of wind and water mills. However, since the industrial revolution the extent and pervasiveness of the role played by technological change has undergone a qualitative change. This rapidly increasing role of technological change was noticed by observers and students of socioeconomic development at the time, but not quite in the same way in which we perceive it nowadays. Nineteenth-century economic historians, for example, could not fail to notice some of the consequences of technological change. They observed the new machines, such as Kay's flying shuttle, the steam engine and the mule. But they assumed that technological progress (or a subset of it) was the cause of the acceleration in economic growth then taking place, paradoxically without trying to explain how these machines in fact contributed to that economic growth. The machine became the icon or symbol of economic growth rather than being an understood component of this process (Mathias, 1984). More careful and attentive observers of the processes of technological change were found at the time among scientists and industrial biographers such as Andrew Ure, Charles Babbage and Samuel Smiles (Mathias, 1984). Among eighteenth- and nineteeth-century economists, Karl Marx and

3

Adam Smith were exceptional. Their work (Marx, 1887; Rosenberg, 1982) combined (in different ways) an interest in the fundamental mechanisms of capitalist society with an analysis of how technological change itself was taking place. Apart from these exceptions the dominant approach was one which recognised the fundamental importance of the new machines as a cause of economic growth but one which could be taken for granted and did not need to be explained (Mathias, 1984). Although this point of view appears nowadays too narrow, it has been very persistent amongst economists and economic historians. Related to this point of view are a number of ideas and concepts which for a long time have been underlying economic treatments of technological change. Thus, for example, technology has been regarded as *exogenous* to the economic system. On this assumption the generation of new technologies is seen as independent of economic factors. On the other hand the economic effects of technology, for example its contribution to economic growth, can be considerable, resulting from continuous decreases in unit costs and the opening up of markets for new products.

Only relatively recently this exogenous image of technological progress has started to change. Schumpeter (1928, 1935, 1939, 1942) and Kuznets (1930, 1953) were among the first economists to emphasise the importance of new products as stimuli to economic growth. Most changes in the analysis of technological change, however, took place after the Second World War. An important stimulus was the great interest in economic growth and the attempt by economists to evaluate its causes. These studies (see Chapter 6) helped both to reestablish the importance of technological change as a very important source of economic growth and to show the limits of the existing approach to the analysis of technological change. These limits became more evident when, after a period of great faith in science and of unprecedented growth in R & D expenditures following the Second World War, an attitude of greater scepticism toward science became common among policymakers in industrialised countries. It was found that there was no direct correlation between R & D spending and national rates of economic growth (implying that the linkages were more complex than had been imagined).

The 1960s and 1970s were characterised by the emergence of *science policy*: the explicit attempt by the state to channel R & D more specifically toward economic and social objectives. If R & D could not be expected to lead automatically to economic growth then the mechanisms according to which scientific and technological inputs could be utilised by the economic system had to be understood far more precisely. This could not be done within the existing approach, mainly based on neoclassical economics, and thus emerged a series of studies of innovation which were not based on a specific theoretical framework. These studies with their much greater attention to the internal features of the innovation process represented a 'natural history' approach to the study of innovation and were a necessary requirement for the construction of a different theoretical framework. Thus, those studies of innovation and of technological change which were aimed not only at analysing some of their effects, but also their origin and the internal structure of the processes, are relatively recent, beginning only in the late 1950s.

From the beginning, studies of innovation and of technological change did not find a precise location within existing disciplines. Important contributions have been made from within a number of different branches of economics, such as production theory, industrial economics, labour economics, and macroeconomics; from within sociology, in particular industrial sociology and the sociology of science; from economic history; from political science, particularly the study of science policy; and latterly from research on the nature of management. Despite this great variety of possible approaches, some common features tend to emerge which suggest some underlying integrity to the process of technological change.

For example it is common to distinguish between product and process innovation. The separability of the two types of innovations may not be complete and it may depend on the particular industrial sector under study but it is nevertheless a distinction which has a general validity. Likewise it is possible to classify innovations as radical or incremental. There may be innovations which are not easily classified, but the classification itself is of sufficiently general validity to be applicable to innovations in various sectors of manufacturing, in services and so on. Other descriptive categories relevant to innovation and technological change, and applicable across a wide spectrum of socioeconomic circumstances, have emerged within the last twenty years. These concepts will be elaborated later in this book; the point which is worth making here is that in this way technological innovation and technological change have emerged as *processes* with a number of distinctive features, to a certain extent independent of the particular areas of socioeconomic life where they occur. Consequently they have acquired a certain autonomy as fields of study. This is not to imply that technological innovation and technological change are independent of socioeconomic circumstances: On the contrary, we shall argue that the common features of these processes are exhibited in different ways in various environments precisely *because* technological innovation and technological change are largely endogenous to the socioeconomic system. What is implied here is an *analytical* separability of innovation and technological change from socioeconomic processes in order to be able to study better their origin and diffusion as processes of interaction between scientific and technological knowledge and socioeconomic variables.

1.2 TECHNOLOGICAL CHANGE AND INSTITUTIONS

Technological innovation and technological change have been described as processes having a number of autonomous features. However, they always take place in particular institutions. Among these institutions are:

(a) Institutions generating new knowledge (universities, private and Government, R & D laboratories, and so on).
(b) Institutions both generating and using new knowledge as an input for their main activity.
(c) Institutions using technological change 'embodied' in products (hardware) and working practices/techniques.

Among the institutional functions which interact with technological change there are production and distribution of commodities, marketing, allocation of investment resources, political and social regulation, military conflict, household structures and others.

The common feature of different economic systems is the achievement of some coordination of the plans of individual economic actors operating within the system. The means by which this coordination is achieved vary not only between different economic systems (for example, the market and managerial mechanisms in capitalist economies and central planning in 'socialist economies) but also between different types of activity within the same economic system. In capitalist societies different branches of activity are characterised by different mechanisms for coordinating individual economic plans. When products can be traded in markets the price mechanism offers a simple and efficient coordinating device. But many products and services cannot be efficiently traded in markets. For example, one could envisage the production of television receivers being carried out by a myriad of firms, each of which would add a component and then sell the incomplete product to the following firm, but such detailed disaggregation rarely happens. Furthermore, some goods conveniently called public goods, cannot be traded in a market because they can be consumed by many users without preventing other users from consuming them, and sellers are unable to restrict use to people who pay for the services. Technological knowledge is in some respects, such a good and, therefore, markets for technology are very imperfect. The production and distribution of technological knowledge is therefore more frequently coordinated by institutions different from the market, such as the firm or the state. The firm may experience an incentive to internalise the generation of technological knowledge (for example to perform R & D internally rather than buying it from specialised R & D institutions) and the state may judge it to be socially optimal to intervene by giving the innovator a temporary monopoly in exchange for information. This is the economic rationale usually given for the patent system. A third situation in which the coordination of individual economic plans is not achieved efficiently through market mechanisms is the joint production of intended and unintended services (externalities). Some externalities lead to hazards, the control of which is difficult by market mechanisms and usually requires state intervention.

Therefore in any capitalist society the coordination of individual economic activities will be carried out by a mixture of institutions the most important of which are the firm, the market and the state, but with other institutions intervening as well. Consequently technological change will be partly 'located' in each of these different institutions. Technological change itself will cause movements of the interinstitutional boundaries that will both cause change in the institutional geography of the economic system and in the location of technological change within it. This implies that, although there is a certain separability of innovation and technological change as processes, they can only be studied in the context of the institutions and of the environments outside the institutions in which they are generated and adopted. From a disciplinary point of view, therefore, the nature of technological innovation is not *deducible* from theories of firms, organisations and markets, but it involves the use of these theories in a fundamental way. Some of these theories will be analysed in this book. The neoclassical and behavioural theories of the firm, the literature on R & D management, organisation theories and business history will all prove relevant for the analysis of technological change.

Business historians have described the remarkable changes which took place in the organisation of industrial enterprises beginning in the second half of the nineteenth century. Small family enterprises, often based on only one activity, either manufacturing or trading, were gradually replaced by large firms in which administration became as important as the main production activity of the firm. The geographical dispersion of their activities, the growing volume of trade, and the processes of backward and forward vertical integration by which manufacturing firms acquired firms producing raw materials and distributing their products led to a specialisation of administrative tasks either on a geographical basis (regional offices) or along functional lines. The *centralised, functional departmentalised* enterprise originated as a consequence of the changes previously mentioned (Chandler, 1962, Chs 1, 2). Beginning in the 1920s the expansion of existing firms' product lines, the quest for new markets and new sources of supply in distant lands, the opening of new markets by developing a wide range of new products (Chandler, 1962, p.42) led to the development of the *decentralised multidivisional enterprise*. These types of firm organisation will be discussed further in Chapters 2 and 3; it is important here to note that technological change has helped to bring about organisational changes. Transport and refrigeration technologies, by allowing the geographical expansion of firms and markets, helped to create more complex administrative tasks and, therefore, to determine the emergence of the new types of organisation. Later the creation of new markets based on new products exerted a similar function in determining the emergence of the decentralised multidivisional structure. What is implied here is not that technological change is *the* cause of the changes in the internal organisation of firms which have taken place in the last century but that it is one of the variables which help to shape business strategies, which in turn lead to changes in business organisation. New boundaries internal to the enterprise have been created due to the increasing complexity of administrative tasks, and technological change has played a role in determining the emergence and the location of these new internal boundaries.

The evolution of the firm from the family type to the decentralised multidivisional enterprise can be interpreted as a progressive specialisation of administrative functions which has led to ever more complex problems of coordination. Coordinating devices such as the general office and formal corporate planning, with an emphasis on defined goals and strategies, were introduced by the firms that pioneered the new forms of organisation between the two world wars but were widely adopted only in the sixties and seventies.

1.3 RESEARCH AND DEVELOPMENT, INNOVATION AND FIRM ORGANISATION

Functions such as sales, production and engineering existed before the changes in firm organisation previously described and the new forms of organisation institutionalised their separation. On the other hand a new function, research and development (R & D), became institutionalised in firms beginning at the end of the nineteenth century. The institutionalisation of R & D has been one of the most central changes in the way in which firms compete and change their technologies (Freeman, 1982). Although the first corporate R & D laboratories were set up in the 1870s in Germany and in the 1890s in the USA, it was only in the period between the two world wars that a significant number of companies, particularly in the electrical and chemical industries, followed this lead. In the industrialised countries as a whole, the real growth period in organised R & D has been the post World War II period. In historical terms then, both formal corporate planning and the experience of managing innovation are still quite recent.

This development of organised R & D in the modern firm has

brought about a dramatic change in the capacity of industry to generate a stream of new and modified products, and to use variations in this capacity as a competitive weapon between firms and between national economies. Most commentators would now accept the prescience of Schumpeter's (1942) observation that the competition posed by new products is fundamentally more important than that of marginal changes in the prices of existing products. This trend then, can be seen as an attempt by firms partly to internalise and control the potential benefits of technological advances, rather than being a victim of them. The specific activities of R & D and innovation, and the more general activity of strategic planning, are therefore attempts to reduce the uncertainty of the future, and to give the firm a greater degree of control over its development in a 'negotiated environment'.

But mere recognition of the need to innovate and the need to plan does not provide the means. The setting up of activities and departments called 'R & D' and 'Planning' can create more problems than it solves, for they are immediately confronted by a vast array of uncertainties and no fixed rules on how to deal with them. Not only does 'R & D' have to be coordinated with all the other activities of the firm but this coordination has to take place in the presence of changes in the environment external to the firm; changes which are themselves important in establishing a strategy for R & D. This problem is illustrated by the simple model of the relationship between the innovating firm and its environment presented in Figure 1.1. This model does not pretend to any analytical status, it is simply an aid to describing the main environmental factors which contribute to innovation decisions, and the principal departments of the firm which participate in them.

Consider first the role of the marketing function in the process of generating and influencing policies toward innovation in a firm. The day-to-day concerns of the marketing staff are with the prices and market shares of their own and their competitors' products. From this information they devise strategies and tactics to discharge their primary responsibility of selling their existing products. An extension of this activity is the provision of information on which future production is based. Clearly, these activities place the marketing personnel in a vital position for observing longer term changes in the patterns of demand in the markets with which they are familiar. The sources of these changes in demand are

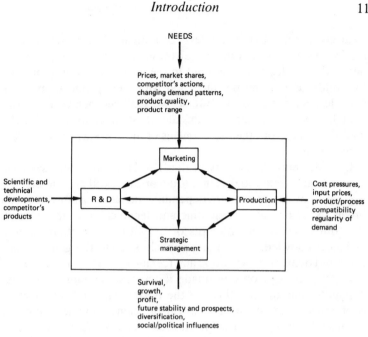

FIGURE 1.1 *A simplified model of the relationship between the innovating firm and its environment*

complex. Some of them appear to originate from changes in the characteristics of the consumers themselves, such as those which stem from increases in per capita incomes or from changes in social and cultural patterns. Other changes might be the result of actions by other producers who change the quality or the range of their products. There is an interaction between changes in the capacity of the supply system, and changes in the basket of needs which find expression as actual demand. This interaction occurs even without the added complication of advertising.

Cause and effect in this interaction is notoriously difficult to disentangle, and the traditional confrontation between notions of consumer and producer sovereignty (Galbraith, 1967) has a close relative in the technical change literature in the antagonism between 'science push' and 'demand pull' theories of innovation. This argument is one which must be resolved at the macro as well as the microeconomic level and is deferred to Chapter 5. What is clear however, is that at the level of individual firm, the monitoring by the marketing function of these changes in the pattern of demand provides a crucial input to the firm's decisions about how to modify its existing products and what new products might be feasible and/or necessary. But, as we shall see below, it is not a straightforward matter to translate these insights into reliable guides to action for other departments of the firm such as the R & D department.

Consider now the activity of the R & D department itself. The scientists and technologists in this department will be connected to a greater or lesser extent, depending on circumstances, to the broad social process of scientific and technical advance. Despite the (often successful) attempts of firms to stop certain items of technical knowledge becoming publicly known, the general character and dynamic of scientific and technical progress is a collective one. Progress can only be made if scientists communicate their results to one another through the normal channels of journals, conferences and so on. The world of science and technology is organised, through its own institutions, across the whole of society, and therefore across firm boundaries. The scientists in a firm, if they are to retain the publicly conferred status of legitimated 'scientist', must continue to participate in this domain, despite the potential conflicts with the needs of the firm to direct their activity. (Some of the problems to which this tension gives rise are discussed in Chapter 4).

Through their interface with the world of science and technology, and through their own efforts to develop and apply knowledge which is relevant to the firms' products or closely related products, scientists and engineers can spontaneously generate ideas for new or improved products. This occurs to some extent independently of any attempt which management may make to direct and organise their scientific resources. Indeed there is evidence from a number of studies to suggest that many of the important scientific ideas in the life of an innovation come from outside the innovating company, via these channels of professional scientific communication. (Gibbons and Johnston, 1974).

By considering the activities of Marketing and R & D at the most general level then, it becomes clear that, despite working in the same company, staff in these areas will tend to view the company's actual and potential products from rather different perspectives. Both will have some view of changes to the product range as being some compromise between demand and possibility. But whereas the marketing perspective will be more likely to centre on need, demand, competition and market size and share, the R & D perspective will tend to centre on technical improvement which is justified 'in its own terms'. Clearly, the most important management function is the organisation of the manner in which these perspectives are made complementary rather than contradictory. The details of this complex process are discussed later in the book. For the moment we are concerned with a general overview of the various actors and their roles.

We turn next to the personnel concerned with the production process. Innovation is not simply a matter for an organised dialogue between marketers and scientists. Any change to a product, no matter how trivial, can have consequences for the way in which the product is produced. There may be a need to change machinery, working practices, materials, sequence of operations and so on. In the case of major innovations which need investment in new plant and equipment, it is clear that the commercial viability of the innovation cannot even be assessed without the assistance of production engineers. Clearly then, the production function in the firm also has a central role to play in the innovation process. As well as contributing to the assessment and development of projects which originate from marketing or R & D, staff in this area will, through their concern to improve efficiency, generate needs of their own which may give rise to innovations which improve the manufacturing processes and product.

The coordination of relationships between these three departments is affected by a variety of influences ranging from the personalities of the individuals concerned, the size of the units and the geographical proximity or distance of the participants, to the nature of the market environment and the range of technical opportunities. But there is in many firms some element of strategic management which structures this process. It does this at the most general level through the budgets it allocates to the three departments. Strategic management may also influence the interactions between the three departments by means of its 'corporate plan', however explicit or otherwise that plan is. For example, consider two firms which both make agricultural tractors and compete directly with each other in that market. One may be essentially a vehicle-producing firm which makes tractors in order to best utilise its automotive expertise. The other may be an agricultural machinery supplier which makes tractors because they are the core product of agricultural machinery. Without anything being necessarily written down on the subject, the engineers and marketers in these two firms may have significantly different views on what are the 'appropriate' innovative projects to think about. Decisions on such broad questions as these, the level of diversification of a firm across quite large product group boundaries, are related to very-long-term judgements about the survival and growth of the firm. They take into account all manner of influences ranging from the possible economic climate and the effects of political developments down to the more obvious but nonetheless difficult issues of the likely competitive threats and the types of response which are available.

As the previous considerations have shown, the institutionalisation of R & D has changed but it has not solved the problem of how to innovate. However, the institutionalisation of R & D has had a number of important consequences for the way in which we think about technological change. First, we can no longer consider inventions and innovations as exogenous to the economic system. Resources have to be allocated to particular R & D projects and, therefore, the direction of inventive and innovative activity is decided from the very outset based on economic criteria. Second, since R & D generates new knowledge and new uncertainty, its existence is incompatible with the assumptions of perfect information and equal technology contained in the neoclassical theory of the firm. Third, the fact that firms choose between R & D projects, implies that there is a spectrum of R & D strategies which have become important components of overall firm strategy.

1.4   STATIC AND DYNAMIC PERSPECTIVES

Research and development and technological innovation continuously create new knowledge and uncertainty. Technological innovation is therefore a phenomenon which is particularly difficult to describe in terms of an economic system at equilibrium continuously adjusting to small disturbances. The concept of an economic system at equilibrium is however one of the fundamental assumptions of neoclassical economics and therefore one which most economists would not easily abandon. However, a number of authors have pointed out that the use of equilibrium analysis alone might be inadequate at least to describe some phenomena. Schumpeter pointed out that long-term growth in a capitalist economy is much more dependent on the creative destruction of innovations shaking and redefining a pre-existing equilibrium than on the smooth working of the equilibrium itself. In other words, the *static efficiency* of a system that at 'every given point in time fully utilises its possibilities' may in the long run be inferior to the *dynamic efficiency* of a system which allocates part of its resources to the generation of new knowledge. (Schumpeter, 1942). The essential difference between static and dynamic efficiency is that the first is achieved by working within a given set of initial conditions while the second implies the creation of new initial conditions (Klein, 1977, p.12). Clearly the extent to which initial conditions will be changed by a stream of innovations will depend on how radical the innovations are. An incremental innovation such as a slight improvement in the fuel efficiency of a motor car or in the properties of a given polymeric material will obviously require a less drastic re-definition of the initial conditions than did the introduction of the aircraft or the computer. The difference between static and dynamic efficiency may then, at least in some situations, be more one of degree than that between two incompatible alternatives. Nevertheless the existence of these two types of efficiency implies that it is not possible to analyse a situation characterised by dynamic efficiency in terms of the variables and parameters required to describe static efficiency. The tension between static and dynamic modes of analysis is particularly important in the case of technological change, in which the initial conditions of the system are always being redefined.

1.5 THE PROBLEM OF AGGREGATION

Technological innovation is a vital component in the performance of firms and of other organisations. That is not to say that the only effect of innovation is to make firms perform better. The greater efficiency that firms achieve as a consequence of various types of innovation is only an intermediate step toward the achievement of economic growth and economic development. In other words the microeconomic efficiency at the level of the firm is related to the macroeconomic efficiency at the level of national economy. But this leads us to the analytical problem of aggregation; the reconstruction of the whole socioeconomic system by means of some combination of its parts. The types of combinations that will be applicable to specific theoretical problems are not known in general and it cannot be assumed that aggregation will be additive, or that the whole will be the sum of its parts. The assumption that the whole *is* the sum of the parts (for example Solow, 1957; Dennison, 1969) depends on some kind of qualitative similarity between the whole and the parts and amounts to what Klein calls the *fallacy of composition* (Klein, 1977, p.20). Differences between various individual firms and between the behaviour of the firm and of the economy at higher levels of aggregation (for example industrial sector, national economy, and so on) are necessary to preserve some degree of *macrostability*. However, macrostability, defined as an overall dynamic process in which the advances come about quite regularly is incompatible with *microstability*, defined as the stability associated with an unchanging environment (Klein, 1977). Therefore great diversity of firm behaviour is required to produce regular advances at the macroeconomic level.

An important issue addressed in this book is the role which may be played by technological innovation not only in determining the internal efficiency of firms and institutions but in determining the combinations of these economic units to form more aggregated levels of analysis. Thus for example, interrelated innovations or common technologies underlying different industrial sectors can have profound implications for the combinations of individual firms within them. Different levels of aggregation will exist with respect to innovations as they exist with respect to economic units. Thus some types of innovation will constitute a common technological basis for a large number of firms or industries. A potential example is the case of microelectronics (see Chapter 7).

In addition some regularities have been observed in the time patterns of development of innovations. In the evolution of new technologies an initial period characterised by a great multiplicity of product designs is followed by the emergence of a *dominant design* (Abernathy, Utterback, 1975). A similar concept, implying that the multiplicity of potential approaches which may exist at the beginning of the evolution of a technology is gradually replaced by a convergence on a common approach, is captured by Nelson's and Winter's concept of *technological regimes* and *natural trajectories* (Nelson, Winter, 1977), by Sahal's *technological guideposts* (Sahal, 1981) and by Dosi's *technological paradigms* (Dosi, 1982). The specific content of these concepts will be discussed later but here it is important to observe that they may provide common constraints for the technological behaviour of very large numbers of firms and therefore represent a higher level of aggregation of innovations. It is, in other words, possible to envisage two analytically separate but interacting domains, that of economic activities and that of technological activities, each having different levels of aggregation. Further examples of interaction can be provided by the relationship between market structure in the economic domain and technological innovation in the technological domain, or the relationship between structural change, seen as a change in the composition of economic activities, and the emergence of a set of interrelated innovations. These problems will be analysed in the book in order of increasing levels of aggregation starting from a discussion of technological change at the level of the firm.

1.6 SOME METHODOLOGICAL QUESTIONS

Before moving, in Chapters 2, 3 and 4, to the innovation process itself, it is important to consider the methodological implications of the 'aggregation problem'. There is an unfortunate division in the literature on innovation between analytical work and prescriptive work. The analytical work is concerned with such questions as: Why do firms innovate? What are the stimuli? What are the effects on productivity, on market structure, on employment, on economic growth? What are the implications for theory of the firm? What is the relationship between scientific advance and need? Are the returns to innovative behaviour adequately distributed? Examples of this analytical work are Stoneman (1983), Freeman (1982), Kay (1979), Heertje (1977). These questions, which are discussed throughout this book, can be regarded as classical social science questions. They are concerned with objective patterns in the behaviour of various aggregates of economic and social actors.

The prescriptive work, taking some answers to the analytical work as given, asks questions of the general form: How can innovation be better managed and controlled? What is the 'right' way to organise R & D? to select R & D projects? to cultivate an innovative ethos? and so on. (See Twiss, 1980). There is a major problem with this persistence of different approaches in the literature concerning the consistency between them. Clearly, the social scientist studying the patterns of technical change in society at large ought to be able to relate, albeit in a mediated way, the explanations of those patterns to the actual decision processes of the individual economic agents who contribute to those patterns. On the other hand, the agents, assuming that they wish their decisions to be rational is some sense, may be expected to attempt to relate the decisions to observed patterns of a more aggregated nature.

In its broadest sense, this consistency between individual and general modes of explanation is a strategic objective of all social science. In economics it is found in the need to make microeconomics consistent with macroeconomics. In sociology it is found in the attempt to make connections between macro concepts such as class, status and power, and accounts of meaning, identity, consciousness and action. In discussing technical change both of these types of consistency are relevant objectives. There are of course limits to how far such consistency can be achieved. The social sciences are not concerned with natural but with social processes. Natural processes are limited in their 'knowability' by some fundamental principles and by probability theory, but they are essentially closed and predictable in a way that is not true of social processes. The latter are intrinsically more open and evolutionary and they are themselves changed by virtue of becoming an object of knowledge.

Our problem is to have an account of innovation in which we achieve an appropriate level of consistency between two views of the process. On the one hand there is the observer's aggregate view: How can innovations be understood in terms of some broader theory of the process of technical change? On the other hand there is the innovator's own view of which aggregate factors are relevant inputs to decision-making. Both traditions in the literature have tended to evade this problem. The analytical tradition has tended to treat innovation as a profit-maximising decision but to emphasise the crucial factor of uncertainty as a result of the variety of non-economic (in the strict sense) factors which impinge on innovation. In other words it is assumed that decisions can be 'subjectively' rational within the limits of the firm's knowledge and *ex ante*, but that such decisions are not necessarily 'objectively' rational with respect to all knowledge or *ex post*. (See Kay (1979) pp. 36–40 for a fuller discussion of this topic.)

The prescriptive tradition tends to avoid the problem altogether. It typically describes what are are thought to be 'best practice' techniques for the management of innovation, for the creation of ideas, for the control of R & D programmes, and so on, but in general it does not attempt to estimate the actual efficacy of these techniques in an objective way. The tone of the literature is in general one of advocacy, suggesting that the use of appropriate management techniques will at least improve the probability of innovative success if not guarantee it. What the prescriptive view tends to obscure is that innovation is a process in which there must be losers as well as winners. If all the firms use the 'best' techniques there will still be losers. There is an irreduceable element of uncertainty in innovation which cannot be 'managed out'. Nevertheless both traditions contain valuable insights into the innovation process and the broader process of technical change. In what follows we shall draw on both literatures, though the intention will be primarily to develop the analytical tradition.

1.7 TECHNOLOGICAL CHANGE AND PUBLIC POLICY

It has already been noted that the coordination of individual economic activities takes place through a variety of institutions, of which the firm, the market and the state are the most important. The emphasis so far in this chapter has been on the firm and on the market. The state is, however, intimately involved in many aspects of technological change. For example state intervention is justified in some circumstances by the existence of public goods and of the hazards which may be a by-product of technical change. The literature frequently distinguishes different types of state intervention which can be divided into *sponsorship* and *regulation*. The first type occurs when, due for example to market imperfections, certain types of goods or services cannot be provided spontaneously by the market. Basic research, education and military technologies are relevant examples in which the state acts as a sponsor. The second type of state intervention takes place when hazards or other unwanted effects are externalities jointly produced with the main services of a given technology and these externalities cannot be easily eliminated or minimised within the market mechanism itself. These considerations provide an *economic* rationale for state intervention in some activities which are fundamental components of technological change.

However, the role and functions played by the state and by a number of non-state public institutions are due not only to economic motivations but to political motivations as well. Changes in technology can have important implications for the balance of power amongst different groups in society. To give just a few examples, the possibility that new technologies have to displace jobs has obvious implications for workers and trade unions; geographically diversified patterns of diffusion of technologies can lead to the decline of some regions and nations; while differential access to new technologies can have profound military implications. These problems are largely political in nature and therefore frequently result in the intervention of the state as mediator between the interests of different groups. In the case of differential roles of technical change between nations, the political and economic rationales for state involvement become closely intertwined. The sponsorship of technical change by the nation state may be an element in a national industrial and economic policy which is aimed at defending a nation's economic position. Representation of the interests of various social groups by the state has, however, not been the only way in which controversies about the political impact of technological development have occurred. Many pressure groups concerned with technological issues such as the protection of the environment; weapons proliferation and the diffusion of nuclear power have sprung up in recent years. These new pressure groups act as institutions intermediate between the public and the state, in somewhat similar positions to institutions such as the unions which, as part of their institutional goals, have had a longstanding interest in technology. These institutions provide a degree of *vertical pluralism* (as opposed to the horizontal pluralism of political parties, Hodgson, 1984, p.163) in technological choices and are thus of great importance in a field in which the need to maintain a balance between expert judgement and political representation may not be entirely satisfied by a central government. Not least among the effects of attempts by central government and intermediate institutions to influence the rate, direction and consequences of technical change is the greater dissemination of information about technological developments to the public at large. This process has undoubtedly increased the degree of visibility and politicalisation of some technical decisions. In this sense technology is one more aspect of political life and therefore one in which the democratic representation of citizens' rights is now seen by many to be as fundamental as it is in education, health care, and so on. The issues of what technologies are going to be developed, how and with what consequences for whom constitute what is sometimes described as the problem of the 'control of technology' (Elliott and Elliott, 1976). The control of technology can here be defined as the political distribution of decision-making about technological developments. It therefore includes sponsorship as well as regulation. This idea of 'control' must be clearly distinguished from the more restricted concept of control as the regulation of technology and prevention of unwanted side-effects such as hazards. The broader concept of control of technology includes the more restricted one.

Sponsorship and control, accountability and public representation are not just conceptual issues. The decision-making and implementation activities involved in the translation of these concepts into practice are carried out by particular institutions and by means of particular policies.

As the exogenous model of the effects of science on economic growth has diminished in credibility, and the evidence about the intimate economic linkages of technology has grown, governments have become more and more involved in the development of science and technology policy. (OECD, 1971). This has lead to considerable discussion and some diversity of behaviour in designing institutions to intervene in the process of technological change, for purposes both of promotion and control. (Rothwell and Zegfeld, 1981). A significant component of the apparatus of public policy and administration in industrial societies is therefore specifically designed to influence the interactions between firms, markets and other institutions in the conduct of technical change. It is therefore not surprising that some political scientists have sought to analyse these phenomena (Williams, 1971, 1983/84, Gummett, 1980). These issues are examined in Part III of this book.

A range of issues connected to technological change have been raised in this chapter, starting from those related to the firm and proceeding to the market, to the state and to intermediate institutions. In so doing the chapter has anticipated the structure of the book, which moves from the microeconomic to the macroeconomic level of aggregation and from economic to political considerations.

2 The Firm, Production and Technical Change

Chapter One has introduced some of the reasons for regarding technological change as an important and an increasingly discussed feature of the structure of industrial societies. However, the pervasive nature of technical change makes it difficult to explain its role in detail. The disciplines and techniques of all the social sciences are relevant, and for empirical studies there is no substitute for gaining familiarity with particular areas of science and technology. The approach taken in this book, which has been outlined in Chapter One, is to examine technical change as a process which is partially located in each of several broad classes of institution the most important of which are firms, other generators of technical knowledge, markets, and political institutions. Technical change also contributes, along with other factors, to the historical shifts in the boundaries between these institutions, which in turn influence future technical change.

In this chapter we begin to examine in more detail the relations between firms, markets and technical change. This will involve some consideration of neoclassical theory of production and of the firm, of behavioural and managerial theories of the firm, and of their potential for explanation of firms' involvement in technical change. In Chapters 3 and 4 these ideas, together with the descriptive account of the innovative process introduced in Chapter One, will be used to examine some aspects of strategy, organisation and execution in innovation. One objective of the analysis is to achieve some greater interaction between the 'academic' and the 'practical' literature than has been common. However, some separation of these traditions is necessary for the purposes of presenting the arguments. This chapter is primarily concerned with a discussion of the firm and technical change within the discipline of economics.

2.1 THE NEO-CLASSICAL THEORY OF PRODUCTION

Microeconomic theory of production occupies a rather central place in economics. It analyses the relationships between quantities of inputs and outputs in the productive unit, and in so doing makes certain important simplifying assumptions about the nature of firms and markets. It therefore combines a theory of production with an implied theory of the firm and a theory of price, or market mechanism. This combination of functions in the theory of production sometimes leads to some confusion over its real purpose. We shall see below that its presentation of 'the firm' is indeed extremely abstract and lacking in realism. This is sometimes interpreted as a basis for criticisms and for the superiority of 'realist' theories of the firm, which are not wholly justified. The abstraction involved in creating a theory of production is a legitimate procedure if the resulting theory performs its primary purpose, which is to model production and market behaviour. Thus many writers have pointed out that the theory is a theory of *markets* first and foremost (Machlup, 1967), and the assumptions made about firms are those found to be logically required as axioms to support this theory of markets. (See also Moss, 1981, Chapter 1, and Rosegger, 1980, Chapter 3). The neo-classical theory is an important starting point for our discussion because some important work on technical change (discussed in later chapters) has taken the route of developing the analytical apparatus of the microeconomic theory of production; and because it does provide a starting-point for later theories of the firm.

The remainder of this section gives a brief summary of neoclassical production theory. More detailed accounts which are oriented towards the discussion of technical change can be found in Stoneman (1983), Rosegger (1980) and Heertje (1977).

The following assumptions are normally regarded as the framework for the theory:

1. The firm produces one homogeneous product.
2. The firm has perfect information concerning inputs and outputs.
3. The firm is a price-taker: its decisions to buy or sell inputs or outputs will not affect the prices of those commodities.
4. Demand and supply are in equilibrium in relevant markets.

5. The firm is owner-managed.
6. The firm maximises profits.

Production in the firm is viewed as the combination of inputs or 'factors of production', the most common of which are capital and labour, in order to produce output. At any time there is a given level of technology which determines the techniques available for production. This in turn determines the maximum level of output which can be obtained from a given level of inputs. A technique is therefore effectively defined as a particular combination of factors of production. Among the available techniques the firm will choose the one which, given existing factor prices, minimises total production costs.

The assumptions and relationships described above are presented in analytical form in the *production function*, which is conventionally given the following form:

$$Q = f(K,L)$$

where Q is output, K capital and L labour. The production function can be represented graphically as a series of isoquants; curves corresponding to the constant output obtainable by the infinite number of available combinations of factors of production (techniques). Depending on the convention adopted the two isoquants Q_1 and Q_2 illustrated in Fig. 2.1 can either represent an increase in output ($Q_2 < Q_1$) or an increasing efficiency in the production of a constant output ($Q_1 = Q_2$). According to the latter convention which is used here, Q_2 would be obtained by using smaller quantities of both inputs.

At a given time the firm will choose from among the infinite number of techniques available on the most advanced production function (Q_2 in Fig. 2.1) that which minimises production costs. It can be shown that this technique is represented by the point at which the straight line having a slope equal to the ratio of the prices of the factors of production is tangential to the isoquant (Point C in Fig. 2.1). While all the points on Q_2 correspond to the maximum technological efficiency, only C corresponds to the maximum economic efficiency.

Technical change in the neoclassical theory of production takes place in the form of shifts of the production function towards the origin. For example, in Fig. 2.1 Q_2 corresponds to a more

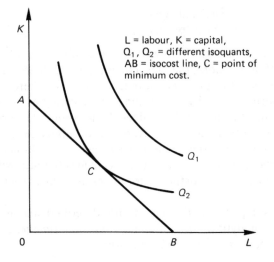

FIGURE 2.1   *The production function*

advanced technology than Q_1. However, a movement of the production function towards the origin can follow a number of paths. In Fig. 2.2, Q_2, Q_3 and Q_4 all correspond to more advanced technologies than Q_1 but the technological change has been qualitatively different in the three cases: at constant factor prices Q_2 uses the same combination of inputs as Q_1, Q_3 uses proportionately more capital and Q_4 uses proportionately more labour. $Q_1 \rightarrow Q_2$ is an example of *neutral* technical change, $Q_1 \rightarrow Q_3$ of *labour saving* technical change and $Q_1 \rightarrow Q_4$ of *capital saving* technical change. Non-neutral types of technical change are called *biased*. (The concept of biased technical change is discussed in more detail in Chapter 5).

If at a given time there is a change in the ratio of factor prices, represented graphically by a change in the slope of the straight line AB (Fig. 2.1), a rational firm would choose the technique which minimises costs in the new situation (C in Fig. 2.3). The possibility of substituting one factor of production for another is, however, not infinite. The convex shape of the isoquants means that, for example, as more capital replaces labour, the process of substitution becomes increasingly more difficult. In this case a change in the economic environment would not have caused any technologi-

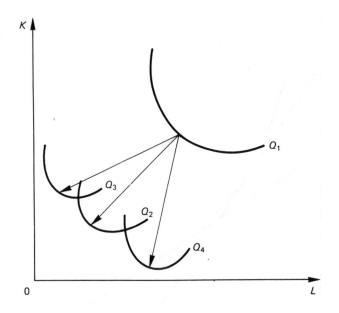

FIGURE 2.2   *Neutral ($Q_1 \rightarrow Q_2$) and biased ($Q_1 \rightarrow Q_3$ and $Q_1 \rightarrow Q_4$) types of technological change*

cal change. However, it is conceivable, and indeed confirmed by historical evidence, that existing techniques could be modified or new techniques invented to use proportionately less of the most expensive input. In other words a change in the economic environment could either cause a change in the technique chosen among the presently available ones or *induce* an innovation, which could produce a more suitable set of techniques.

Within the neoclassical theory of production the distinction between these two alternatives has been possible only under some restrictive assumptions. Furthermore the mechanisms according to which induced innovations could be generated are not yet clearly understood. Discussion of the theory of induced innovation is deferred until Chapter 5. For the moment it is sufficient to note that a shift in the production function could take place either as a result of independent (exogenous) changes in the state of knowledge or as a result of changes in the economic environment. We shall also return to this point in discussions of science push/

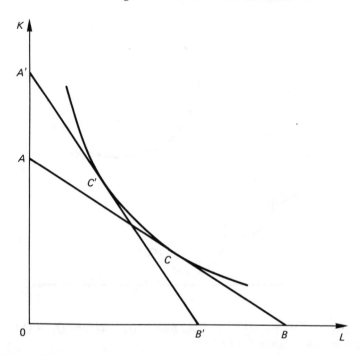

FIGURE 2.3   *Change of the economically most efficient technique due to a change in factor prices*

demand pull theories of innovation and of economic growth accounting in Chapter 5. In summary then, the theory shows each firm as facing a production function and making a choice of technique on the basis of prevailing factor prices. Changes in the spectrum of available techniques are exogenous.

There are clearly some limitations to the neoclassical theory of production which affect its value as a framework for explaining technical change. The following are the most important ones:

1. Only labour and capital are incorporated as factors of production. This is obviously an oversimplification but it is not a fundamental difficulty. The inclusion of more factors of production does not present any conceptual difficulty although it makes the application of the production function analytically more complicated.

2. The presence of infinite techniques at a given level of tech-

nology is rather unrealistic: real-life situations often imply a choice between a restricted number of options.
3. The substitutability of labour and capital is sometimes limited when, for example, they are bought in lumpy, indivisible units.
4. Only changes in process technology can be described by means of the production function. Although it is true that the product technology of one industry is sometimes the process technology of another industry this is not necessarily the case with final goods.
5. Only cost-reducing improvements can be described by the production function. Improvements in performance or the appearance of new services find no place in the neoclassical theory of production.

2.2 POST NEO-CLASSICAL THEORY OF THE FIRM

The previous section has made clear that, whilst neoclassical production theory allows certain clear but abstract definitions of technical change, it makes assumptions about the nature of firms which lack realism both in general terms and in terms of the process of technical change. Since we are concerned to examine the interaction of firms and markets in the process of technical change we cannot be content with an approach which reduces firms to epiphenomena of markets, and realism must be an important objective in the accounts of firms which we use. This section reviews some of the principal results of 'realist' theories of the firm. This provides a basis for a later section which explores in more detail the scope for treatment of technical change within such realist theories.

Hay (1983) argues that two major historical changes in the nature of firms have helped to precipitate new developments in the theory of the firm. Firstly, the growth in the number of very large industrial firms during most of this century has made it increasingly difficult to rely on a theory which presents the firm as atomistic and small in relation to its markets. Secondly, since the seminal work of Berle and Means (1932), there has been extensive discussion of the so-called 'divorce of ownership and control' in industrial firms. The significance of this second point is that it establishes *the scope of managerial action and its motivations* as a central theoretical concern in explaining the behaviour of firms, thus providing an alternative point of reference to market phenomena alone. Since most of the theories of the firm since then have given some important role to managerial behaviour it is worth briefly examining the 'ownership/control' arguments, before turning to the theories themselves.

The central point of the ownership/control argument is the dispersion of shareholding in companies and the consequent difficulty for shareholders to act as a group to determine or even successfully monitor the activities of managers. The argument has a practical and a theoretical dimension. In practical terms, quite small shareholdings do allow some basis for control (Beed, 1966), but since senior managers often possess holdings on such a scale, they are best placed to take advantage of this possibility. Furthermore this is coupled to their unique position as a group capable of proposing policies for the firm which are very likely to gain them the support of other small shareholders. Shareholding by large institutions such as pension funds might be thought to be a counterweight to this autonomy of managers, but Hay (1983) argues that in the UK, institutions do not appear to choose the option of intervening in the affairs of companies, but simply to buy and sell shares according to their assessments of performance. (In other countries such as Germany there may be institutional representation on the management board, but since these representatives will also be managers rather than owners, it could be seen as joint management rather than owner-control).

The theoretical dimension of the argument is based on the uncertainties surrounding both managerial plans for the firm and the goals of the shareholders themselves. Managers' autonomy is at least partly due to their greater knowledge of the firm and of the environment in which it operates. Naturally this autonomy will be limited by the requirement to satisfy shareholders' expectations in terms of some overall criterion of performance based on profit rates, liquidity, rates of growth and so on. In essence the knowledge gap between managers and shareholders will be one of the sources of managers' autonomy. This autonomy will therefore be more limited when shareholders have a greater knowledge of the firm's business or some particular reason for intervening directly in the affairs of the firm, as happened in the case of Distillers Limited, the company that marketed Thalidomide in Britain. A few concerned small shareholders organised a meeting of the shareholders which instructed management to settle with the families of the Thalidomide victims. At that point, legal action had dragged on for more than ten years showing no sign of reaching a just settlement, and the company's management appeared to have been resisting such a settlement. Nevertheless the ownership/control debate established that managerial action is an important influence on firm behaviour, and that such action is logically and practically distinct from that implied by the owner-entrepreneur model in traditional theory. From this we are led naturally to the question of what motives will underlie managerial action.

First, however it is worth noting that in addition to that aspect of the ownership/control debate which is concerned with the locus of control, many of the contributors to the discussions in the 1950s and 1960s were concerned with what they saw as profound implications of the supposed divorce of ownership and control for the long-term evolution of industrial societies. Tomlinson (1982) provides a succinct summary and criticism of these aspects of the literature. He argues that some writers, represented most clearly by Crosland (1952) saw the emergence of 'managerialism' as a fundamental break from previous forms of capitalism, and offered the prospect of socialist-oriented political change without the confrontation with a capitalist class implied in classical Marxism. The opposing school, represented by such examples as Barrett-Brown (1958) argued that various practical and ideological considerations meant that the supposedly autonomous managers still represented the interests of the capitalist class. The primary concern of these writers was to defend broad intellectual and political stances within discussions of the firm's evolution. More recently, however, most writers in this area have seen the nature of firm behaviour as a topic deserving the development of its own body of theory before attempting linkage to more aggregate levels of theory.

Tomlinson, for example (1982), argues that earlier writers were wrong to continue to put the shareholders in the position of *owners*, and then to insist on measuring their power against that of the managers. He argues that since the corporation is itself a legal personality and is not formally owned by shareholders, it is more appropriate to regard the *physical* possession of the firm by its managers as a reference point for discussions of control than the question of ownership in the strict legal sense. In this perspective control of the firm is not reduced to an argument about what a group of individuals do or do not 'represent', but depends on an analysis of 'a series of practices which construct particular loci of decision-making with particular modes of calculation and thus with a determinate set of possible outcomes'. Tomlinson sees earlier writers, and in particular socialists of various kinds from Marx to Barrett-Brown, as being limited by a lack of a theory of the firm. The result is that they deduce firm behaviour *only* from various systemic properties of capitalism. Tomlinson is not unconcerned with these latter issues, but argues for firms to be examined as complex, internally differentiated institutions. In so doing Tomlinson is simply putting within his own theoretical framework those issues which have been included in more recent discussions of the firm and managerial behaviour in the sociological and management literature. The core of these issues, as mentioned above, is the analysis of managerial motivations, and it is these to which we now turn.

2.3   MANAGERIAL MOTIVATION

Since the work of Marris (1964) it has become conventional to discuss managerial motivation under two headings. First are a group of sociological and psychological variables such as attitudes to achievement power, status, and scope for creativity in the organisation, all of which are seen as directly influenced by the economic success of the organisation in which the manager works, but in a rather complex manner. The second heading concerns the financial remuneration of the manager, possibly comprising performance-related bonuses and stock-options as well as basic salary. These also are clearly conditioned in some way by the nature and performance of the organisation. However, operating to some extent in a contrary direction, is the desire for security both for the individual manager and for the organisation as such. Marris's analysis of these variables and their interactions leads him to suggest that managers will make *growth* of the firm their most important objective, but that this growth will be subject to constraints. The form of this constraint is the threat of take-over. It is argued that if growth is 'too' rapid, the valuation ratio of the firm (the ratio of market to book values of assets) can fall to a point which invites a take-over bid. It is the mechanism which lowers the valuation ratio which is particularly interesting. Marris argues that there is a trade-off between growth and profitability and that it is the lower profitability associated with higher growth which lowers the valuation ratio. The reasons for proposing this trade-off are: first, that growth in existing markets must involve reducing margins to increase share; second, that growth in new markets involves a move away from existing activities, implying higher risks and higher costs; and third, that rapid growth requires the recruitment and training of new managers who do not immediately possess the specific skills necessary for the best management of the firm.

The third point is a development of the basic theoretical insight in another of the key contributions to this literature, that of Penrose. Penrose argues (1980) that a firm is a bundle of physical and human resources and that its growth is fundamentally conditioned by the managerial resources it possesses and their expression as managerial services. They create growth in that they identify and exploit new opportunities for diversification, and increases in efficiency. But they also constrain growth in that they create an internal limit on the rate at which new managerial resources can be acquired and trained to manage the new opportunities.

Thus the combination of the theories of Penrose and Marris provides a powerful managerial theory of firm behaviour, which predicts that firms will pursue growth and size as intrinsic objectives, and that profit, sales, market share, and other goals will be seen as instruments to achieve those ends: all subject to a constraint framed in terms of the continued existence either of the firm itself or of the managerial team which control the firm. This theory permits a number of strategies for testing it, which principally involve studying the relationships between growth, size and managerial remuneration, and between owner-control, managerial control and firm performance. Hay (1983) in his review of this literature concludes that this empirical work has not given very great support to the managerial theory of the firm, but that its intrinsic plausibility suggests that it will (with some justification) continue to influence research. He argues (1983, p. 72) that an appropriate research strategy is to disaggregate the managerial theory into more detailed components for individual study. This implies looking at the relationship between managerial motives and firm *structure*, thus introducing the internal differentiation which exists in the management resources of any particular firm. It is this aspect to which we now turn.

2.4   FIRM STRUCTURE

There are a number of different approaches to the question of structure to be found in the literature. Three are particularly relevant for our purposes. These are the transaction costs approach associated with the later work of Williamson; the 'behavioural' approach stemming from the work of Simon and of Cyert and March, and the historical approach, particularly that associated with the work of Chandler and mentioned in Chapter 1, but also associated with the earlier work of Williamson.

The transaction costs approach is designed to examine the determinants of the boundaries between market forms of coordination (for example, buying in components or services) and administrative or hierarchical forms of coordination (by doing it yourself). Williamson's (1975) thesis is that the principal influence on this boundary is the attempted minimisation of the transaction costs associated with the conduct of a particular set of connected economic activities. Thus when managers perceive an opportunity to reduce their transaction costs by converting a market, or external contracting relationship, into an administrative or internally integrated relationship (or *vice versa*), then they will attempt to realise that opportunity. Certain factors are seen by Williamson as being most likely to influence change in the direction of hierarchies rather than markets, for example in the classic case of vertical integration. These are, adapting his terminology somewhat: uncertainty and bounded rationality, small numbers of agents involved (which places a premium on personal contacts and first-mover advantages), lack of candour in transactions, and asymmetry of information between agents. In these circumstances he predicts that associated productive activities will be brought under the same administrative 'roof'.

This approach has some intuitive appeal and contains possibilities for the development of a technical change component such as that attempted by Moss (1981) (discussed later in this chapter). However, the approach also has some fundamental weaknesses. Whitley (1984) has pointed out that the approach requires the assumption of a universal pressure to minimise transaction costs yet it does not specify the source of this pressure nor the particular mechanisms through which managers become aware of it or design responses to it. Furthermore, it *reduces* hierarchies (particular firm structures) to the consequences of changes in transaction cost incentive and does not allow for the differentiation which exists between managerial groups within hierarchies, their potential for conflict with each other, and the consequences of this for the conduct of the firm. These points receive more attention in the second of the three perspectives on firm structure: that of Cyert and March (1963).

Their model is based on differentiating the goals of the firm, and making some particular assumptions about the way decisions are made in connection with those goals. The five major goals are associated with production, inventory control, sales, market share, and profit. It will readily be seen that there are potential conflicts between these goals. For example a high level of inventories may be favourable to sales but unfavourable to profit; a lower inventory level requires a flexible production system whereas production managers prefer stable production goals, and so on. The firm is presented as a coalition of managers bearing variable relationships to these goals and experiencing and responding to conflict between them. The model makes the following assumptions about the way in which they attempt to resolve the conflicts: first, the goals are conceived as aspirations or target levels rather than as maxima or optima, thus the behaviour is described as 'satisficing'. Second: decisions are taken incrementally focusing on one issue at a time, thus reducing some of the potential conflict. Third: there is some degree of 'organisational slack' in the system which permits 'side payments' between sub-units of the firm and allows different levels of efficiency between sub-units. Fourth: standardised operating procedures or 'decision rules' tend to be established within the sub-units of the firm and these allow day-to-day operations to be conducted without perpetual reconsideration of the conflicts between goals and sub-units.

This is clearly a rather different picture of the firm from the efficient 'minimiser of transaction costs' presented in the work of

Williamson discussed above. Firms are seen as rather fluid structures whose ultimate behaviour is a rather unpredictable resultant of a number of interacting components, each of which may relate to environmental variables in different ways. However, the model stops short of explaining how *particular* types of outcome in firm behaviour might be deduced from specified circumstances or boundary conditions and in this sense it is incompletely specified.

It can already be seen from what has been said so far that the attempt to combine discussion of firm structure with managerial motivation does not instantly generate a superior solution. What for example is the result of combining the structure from the Cyert and March model with the motivations and behaviour implied in the Marris model? Does it take us any further in explaining the behaviour of different firms in different circumstances? Such questions remain to be answered by researchers in this field.

However, it does seem likely that progress can be made if the question of explaining *variation* in firm structure and behaviour is taken as a starting point. Comparison of contrasting major 'types' is a traditional method in social science analysis and it can be extremely useful in this context. This has been the approach of those authors who have pointed to major historical patterns in the structure and behaviour of firms: the third of the approaches to be discussed in this section. The central proposition in this approach is that there is a distinction between U-form enterprises which have a pyramidal structure and specialised functional divisions such as production, sales, finance and so on: and M-form enterprises which have a number of divisions based on products or geographical areas. The divisions function with some autonomy with regard to day to day operational matters but are dependent on the top management for investment funds, which are related to scrutiny of their performance. Williamson (1962, 1967) argues that the U-form enterprise exhibits 'control loss' as it gets larger, and as Hay (1983) points out, it is particularly likely to develop the internal conflicts described in the Cyert and March model. Thus the M-form organisation can be seen as an organisational innovation in response to the problems of size, and a formalisation of the fact that some types of decision such as those associated with the level and direction of investment are different in character from those of the operation of a productive unit and merit concentration in a 'strategic' tier of management. Whereas Williamson's analysis concentrates on the analytical basis for the distinction between U- and M-form, Chandler's (1962) account adds other dimensions of analysis together with extensive historical data. The essence of his arguments has already been outlined in Chapter 1. Some additional points can be made here.

Chandler argues that in the late nineteenth and early twentieth century large U-form companies developed in the US in *certain industries*. These industries were ones where the combination of technological factors and market size favoured the forward and backward integration of companies with raw materials or component production and distribution of final products. The industries concerned were mass production of perishables for mass markets (food and drink); mass production of low-priced semi-perishable products for mass markets; mass production of new machinery requiring specialised marketing skills; and high volume producer goods which were technically complex but standardised (such as those in electrical engineering and chemicals). The technological and market factors which facilitated these developments were electric power and changes in production technology (see also Blackburn, Coombs and Green, 1985, on this point), and changes in communication systems and real personal incomes which enlarged the scope of markets. Where these conditions did not exist, firm growth and the development of large U-form organisations were not as rapid according to Chandler.

Chandler then suggests that these large U-form hierarchies accumulated considerable resources, both in investment and managerial skills, and began to seek opportunities for diversification in order to use these resources. Thus they began to encounter the problems entailed in managing diversified companies with a U-form organisation and so the gradual shift to M-form structures emerged. As Whitley (1984) points out, this identification by Chandler of managerial skills as a factor in the diversification and growth process is very close to the arguments of Penrose mentioned earlier in this section.

The connection of this discussion of U-form and M-form structures to the transaction costs approach subsequently developed by Williamson can readily be seen. The processes described by Chandler in the growth of the U-form firms does depend on some notion of a general tendency to minimise total production costs and distribution costs via strategies of vertical integration. This is potentially reconcilable to the notion of minimising transaction costs, but it adds the specific historical factors which are seen as promoting it in some instances and not in others. Some of these factors are technological, and we shall therefore return to them in due course. Chandler's approach has led to considerable further work by business historians on the development of those large firms in Europe which do not seem to have followed the Chandler pattern in quite the same way as those in the US. (See Channon, 1973; Chandler and Daems, 1980; and Hannah 1976).

A further elaboration of the U-form/M-form discussion has been provided by Cowling (1982). He argues that as firms grew in size within a U-form structure there was a general transfer of unrealised profits into the direct or indirect control of middle-level managers. This took the form of overstaffing, high expenditures on non-salary remuneration to middle managers and in fact expenditure on many of the things labelled by Williamson (1964) as 'expense preference' and related to status and power of managers. Cowling argues that this drain on profits was recognised by senior managers as contrary to *their* strategic interests and so the M-form organisation was innovated in order to place operating units under a tighter discipline and restore the level of aggregate profits. Cowling presents data to support this claim, but the relative significance of Williamson's, Chandler's and Cowling's explanations of the development of the M-form organisation has to be considered an open question. All three are theoretically plausible, and are therefore likely to be operative to different extents in different specific historical circumstances.

This compressed review of some of the issues raised in the literature on the theory of the firm allows us now to draw some conclusions of relevance to the analysis of technical change. The post neoclassical literature presents firms as managerial hierarchies which have a considerable degree of autonomy in their action, are internally differentiated, face considerable uncertainty, and are likely to pursue growth as an objective, but via a variety of different strategies, which in turn depend on factors both internal and external to the firms themselves. These strategies will often be directed toward the *creation* of production functions rather than the observance of them (Moss, 1981), as well as toward the creation of markets. In order to explore the formation and implementation of strategies in more detail it would be necessary, ideally, to take into account the sociological attributes of managerial elites with their national variations, the market and technological opportunities and constraints facing firms, and the interactions of the strategies and actions of competing firms. Such a range of tasks is too large for the present discussion, and is more in the nature of research agenda. This is especially so since it has the extremely radical implied effect of turning neoclassical theory on its head and building a theory of markets from a theory of the firm rather than the other way around. The work of Moss (1981) and Earl (1983) are partial contributions in this direction but the goal is still some way off.

However, if the radical task is yet to be achieved, we can nevertheless make some progress on the more modest task of identifying some of the connections between managerial strategies and actions, and the technical opportunities and constraints which are one variable in the picture, and are the primary concern of this book. The work of Chandler has already pointed to the prospect that analysis of variation in managerial behaviour is a fruitful way to uncover the relationship of the behaviour to other variables.

2.5 TECHNICAL CHANGE AND THE THEORY OF THE FIRM

We referred in Chapter One to the relatively recent nature of the institutionalisation of the R & D function in the firm. Only in the period between the two world wars did private R & D laboratories appear in any numbers, and the really major expansion took place after World War II. Several of the authors discussed in the previous section have drawn attention to the historical overlap between this rise in internal R & D and the spread of the M-form organisation (Chandler, 1962; Marris, 1971). Kay (1979) argues that 'the development of R & D as a specialised integrated function in the firm is complementary to the adoption of the highly structured and functionally differentiated M-form'. In essence, the M-form organisation and the use of R & D to create new products share an important underlying component: namely the attempt on the part of the top layer of management in the M-form organisation to negotiate and control some aspects of their environment rather than simply responding to it. To this diagnosis we can also add the conclusion from the preceding section that the ultimate purpose of such actions is the securing of stability and security for the organisation as such, and growth of its activities and resources such that managerial preferences are satisfied. Thus the managerial theories of the firm allow us to introduce technical change as an active component of firm behaviour. It should be noted however, that there are many practical and theoretical details concerning the conduct of technical change which are not resolved or explained merely by adopting a managerial perspective. One example is the explanation of the differing structural locations occupied by R & D departments in M-form organisations. This and other topics are taken up in later chapters.

In the remainder of this chapter we shall examine how the growth of the firm, which has been identified as the prime objective of management, can be made into a more well-specified mechanism by the incorporation of technology. Clearly the purpose of an R & D department is to select and shape particular technologies and use them as components of complex strategies to allow a firm to grow. In that sense the nature of the activity is to expand technological possibilities; in short, to create technology. But it is not an attempt to create *all* possible technologies, nor to solve all conceivable functional problems with new products. We can expect that the use of technology will be constrained by a number of factors. Firstly, both scientific and technical knowledge generate constraints on the possible performance characteristics of technical systems. In other words, some things are simply 'not inventable' given existing orthodoxy in scientific assumptions. This will normally, though not always, preclude any attempt to seek such 'excluded' solutions. Secondly, for any particular firm, a Penrosian argument would lead us to suspect that the specificity of the managerial and technical resources in the firm will place some constraint on the range of technical components which it can incorporate into its growth strategies. Thirdly, we can expect that some technologies will offer different economic or growth incentives, regardless of the relative ease or difficulty of the purely technical problems involved in their development.

These suggestions lead to the hypothesis that technological possibilities are not an infinite set, that within any finite set there are non-technological factors affecting their potential contribution to a firm, and that these structural features of the technical environment fundamentally *constrain* the *direction* of growth which a managerially controlled firm can contemplate or achieve. If technology does indeed structure the direction of growth, then this is an important addition to the literature of the preceding section. The possibility was to some extent anticipated in the arguments of Chandler. Some other contributions can now be introduced which support the notion of technologically directed growth.

The analysis of 'technological imbalances' is one topic which has made an important contribution to our understanding of the relationship between technology and the direction of firm growth. Rosenberg (1969) argues, using a number of historical examples, that technological imbalances frequently occur in those production activities which consist of a number of closely linked steps. If a change in the technology of one step increases the rate of output of that step, then the productivity of the other steps may limit both the exploitation of the benefit in the changed step, and the productivity of the process as a whole. Thus the innovation will have created a new 'bottleneck' in the process. Rosenberg argues that the existence of a bottleneck will focus inventive effort on ways of solving it. Solutions will then create new bottlenecks and further solutions, and so on.

Moss (1981) analyses such processes in a way which makes more explicit use of managerial perspectives on firm behaviour. He suggests that imbalances of the type described by Rosenberg could be more generally expressed in terms of two types. The first of these is the 'focusing mechanism'. In this case the phenomenon is entirely internal to the firm and its characteristics. For example, if one part of the productive structure of a firm is not working at its full capacity (because the other parts of the firm cannot use the total output of the under-utilised section), then there will be an incentive to find a way of expanding that section's output in order to work at full capacity. This may involve expanding capacity in the other activities or it may involve initiating a new activity which will create new demand for the output of the under-utilised section. The fundamental point is that the problem and its potential resolutions are 'localised' around the existing managerial skills and resources of the firm. Thus the argument is essentially Penrosian.

The second version of imbalance identified by Moss is one in which a focusing effect is combined with some environmental factor exterior to the firm which acts as an extra determinant of the solution chosen by the firm. For example, the shortage of a particular input to the productive process, the presence of a particular type of competitive threat (such as a new increment of product quality from a rival), or the existence of backward or forward linkages which could be exploited, all create what both Moss and Rosenberg call 'inducement mechanisms' to change the technology of the firm in particular directions.

This latter argument concerning inducement mechanisms in the context of backward and forward linkages provides the opportunity for a further comment on the analysis of vertical integration by Williamson which was discussed in the preceding section. It will be recalled that Williamson regarded the mechanism which leads firms to engage in strategies of integration as the desire to minimise transaction costs. The existence of these costs was seen as being essentially the result of inadequacies of information resulting from such factors as the lack of candour of the participants in the exchanges between the un-integrated activities. Moss criticises this view, and argues that it is the technological factors determining the two productive activities which will in turn determine the transaction costs of the participants, and not the informational costs. He provides a number of examples (p. 157 to 160) to support this judgement. There is undoubtedly a great deal of substance in this point: in Moss's example of the inducements to integrate steel production and steel rolling it does seem more plausible to see the perishability of hot steel as a better argument for putting the two activities on the same site, rather than the argument that steel rollers find it difficult to enforce contracts for hot steel on steel producers. Most examples of integration, however, might not be so clear cut in their technical aspects as this one, and it seems rather difficult to deny the general importance of information costs when many advances in information technology are aimed at decreasing precisely these costs. While Moss's arguments deserve general support, it is important to register that other factors such as those mentioned by Williamson may also be relevant. Thus, when we say that technology may determine transaction costs and thus possibilities for integration, it would be advisable to interpret this as determination in the sense of 'setting limits around' rather than determination in the sense of exact calculus. Our earlier discussion of Williamson's work made the point that his analysis does not have room for internal structure and conflict within the firm's management in the manner suggested by Cyert and March. The attempt to determine transaction costs by technology alone rather than by information issues may also run the risk of excluding the effects of internal conflicting goals in the firm.

This discussion of technological imbalances then, has illustrated one possible link between technical change and firm behaviour. The arguments proceed from description of specific managerial resources, technological opportunities and constraints, and environmental constraints; and deduce the direction of managerial initiatives, technical changes, and new firm structures and firm/market boundaries.

We shall now examine another related topic which has contributed to the elaboration of the links between technical change and firm behaviour. Many authors have commented on the fact that industries differ substantially with respect to the opportunities they offer for technological change. In the work of Chandler, already mentioned, the role of technology in permitting, or contributing to the growth of firms, is seen to differ between industries. Kay (1979) cites a number of studies which have suggested that higher ratios of R & D to sales in some industries are a reflection of the greater returns to R & D in those particular industries. (For evidence on the dispersion of R & D intensity across industries see Freeman (1982, Ch. 1). Kay's model of the determinants of R & D budgeting at the level of the firm (which is discussed in more detail in the next chapter), makes variation in technological opportunity an important determinant. His empirical data give some support of this view. It is therefore interesting to examine the concept of technological opportunity in a little more detail. We shall use three different methods of exploring this concept here: one based on technologies themselves, a second derived from essentially neoclassical economic logic, and a third incorporating managerial arguments.

Most technical systems can be described in terms of some performance characteristics which are of significance in the conduct of their primary function, and some physical specifications which are defined in terms specific to that technical system. A simplified example can be found in the circuitry of microelectronics: the speed of operation of such circuits is an important performance parameter with which users are very concerned, and this in turn is partly a function of the degree of miniaturisation of the circuits. (The smaller the distance the electrons have to travel the more quickly the circuit operates.) However, there are some physical limits to the degree of miniaturisation possible. So long as this limit is still far off, then we might say that there are still substantial 'technological opportunities' present in this system. For the moment we are ignoring the problems and costs which might impede the exploitation of such opportunites. These can be introduced later in the analysis. For the moment we are concerned only to establish that some opportunities are more significant than others in principle, and that this is at least in part a matter of the technologies themselves. The former is an example of what could be called an 'intensive' technological opportunity, in that it consists of improvements in an existing specification/performance relationship. We can also imagine 'extensive' opportunities. These would exist where a particular technology, with its particular specification performance relationships, has the possibility of being transferred into a large number of other technical systems where it can serve a variety of functions more efficiently than the existing technologies in use. Microcircuitry again affords an obvious example, but there are many others, particularly in the field of synthetic materials.

Discussion of technologies in these terms can be more formalised (Saviotti, Metcalfe, 1984) and can be used as a basis for discussion of non-price competition (Gibbons *et al.*, 1982, and Chapter 6 of this book). What is significant for our discussion in this chapter is that the technological frontier has some 'structure' and that technical reasons exist to expect some strategies for expanding the frontier to be 'easier' than others.

However, ease of expansion of a frontier is a relative concept and therefore needs to be expressed in terms of some unit. The obvious one is consumption of resources, and this leads us to the second way of examining technological opportunities which is derived from orthodox economics. The essential task here is to express some relationship between the cost of achieving some unit of technical change, and the benefit or profit which falls to the firm which makes the change. There are a number of ways of approaching this problem: we shall start from an adaptation of the arguments of Schmookler (1966), which stems from the work of Stoneman (1983). Schmookler's work was concerned with the relationship between patented inventions in capital goods industries and the level of investment in industries using the capital goods. In fact, the major discussions of his argument have occurred in the so-called 'science-push, demand-pull debate', which we defer to Chapter 5. Here we are concerned with the possible use of his initial argument for the purposes of describing variable technological opportunities.

Stoneman (1983) presents the argument in the following form: let

$$G = p.x - c.x - E$$

where G is profit
p is the price of the commodity incorporating the invention
x is the level of output of the invention
$x = sM/p$
s is the market share for the inventor/producer
M is the size of the total market
c is the cost of manufacture of one machine
E is the expected cost of making the invention.

If a mark-up pricing policy is assumed then the expression becomes:

$$G = s.M\,(1 - k) - E$$

where $k = c/p$. Thus profit is positively related to size of market, share of market, and mark-up, and is negatively related to cost of making the invention. (Problems of measuring a unit of technological change are here avoided by using invention as the unit). This simple expression now allows some useful insights into the nature of technological opportunities. If we consider first a fairly radical invention; one which perhaps opens a new market, or serves an existing market with a product based on new principles, then we can see that technological opportunity will be high where s and M are high, and high in relation to E. M and s might be high if the performance or the cost of the product incorporating the invention offer major advantages with respect to previous experience, and E will be low to the extent that the knowledge involved in making the invention is familiar or easily available *to the firm making the invention*. Consider now the case of subsequent improvement inventions which increase the performance or reduce the price incrementally with respect to the initial invention. M and s will depend on the incremental growth rate in performance. This will perhaps increase in the short run as a result of learning, but will diminish in the long run as a result of encountering the intrinsic performance limits of the technology. For these improvement inventions E may also diminish as a result of learning. Thus the intrinsic benefits of the subsequent innovations are analytically related in some way to the speed and extent of diffusion of the first invention. This point is pursued in more detail in the section on diffusion in Chapter 5.

Although the relationship is not exact, it can be seen that the first case discussed above is rather similar to the case of 'extensive' technological opportunities discussed earlier, whereas the second, incremental case has more similarity to the case of 'intensive' technological opportunities discussed above. Stoneman follows Rosenberg (1974) in observing that Schmookler himself did not discuss the possibility that the cost of invention might be variable and related to the varying technological opportunities across industries. Stoneman's further contribution is to extend the initial part of the model above to the case of profit-maximising firms operating in a particular industry. He finds that the cost of invention, or technological opportunity variable, retains its importance. Stoneman does not however discuss the factors which may affect the distribution of technological opportunity in the way attempted above.

One of the results of the above discussion is of further interest as a result of its long-term character. The suggestion was made that where intensive technological opportunity exists, it takes the form of a series of incremental inventions or innovations which improve the relationship between a particular specification and performance variable, or a bundle of such variables. This set of circumstances is very close to that described, from another theoretical perspective, by Nelson and Winter (1977) in their discussion of 'natural trajectories' of technical change. This concept is taken to mean a series of technical changes which have common features, either by virtue of application of the same principle or by virtue of application to the same system, or both. Examples of such trajectories and their significance to the patterns of technical change in the economy as a whole are discussed in Chapters 5, 6 and 7. We note however, at this stage, that this work is another support for the proposition that the 'supply side' of technical change is structured in very important respects. This changes the relative costs of different directions of technical change, and therefore affects incentives operating on alternative strategies of firm growth.

Finally we turn to the managerial theories of firm behaviour and explore what light they can shed on the question of technological opportunities. Firstly, it seems plausible that the Penrosian argument concerning the specificity of managerial resources can be extended to cover the skills, experience and expectations of the scientific personnel working in the R & D department of an innovating firm. Despite the attempts of R & D managers to maintain and extend the skills of their staff (see Chapter 4) it is likely that the nature of their work on particular projects over substantial periods of time will tend to make some aspects of their knowledge and ability in some sense specific to the products and technologies with which they are familiar. These restrictions may still leave them with a large field of alternative directions of further R& D work within their scope, but not an infinite field. The organisation may seek to compensate for any deficiencies by recruiting new personnel with other skills, but their ability to apply that knowledge within the context of the firm will perhaps be limited in the first instance by their lack of familiarity with the new firm and its routines and structures. This is in fact a more general statement of the focusing device proposed by Rosenberg and Moss and discussed above, but extended to the area of specific technical competences. It should be added that technical competences are often relevant to the staff of other functional areas in the firm, as well as within R & D itself. This may be particularly true of marketing functions. These circumstances will constitute a constraint on the technological opportunities facing a firm.

Furthermore, the internal structuring of R & D into groups of workers, groups of skills, groups of product problems, and the consequent sharing of equipment and reinforcement of patterns of work will all tend to strengthen the effect described above. Kay (1979) points out that such structuring of R & D results in part from the synergies which will result from the grouping of R & D activity into sub-areas of activity. In fact, such considerations are frequently incorporated into the practical aspects of project selection by R & D managers (see Chapter 4). These firm-specific aspects of the technological capabilities of the firm lend weight to the observation made above concerning the cost of invention (E) in the Stoneman/Schmookler model. The cost will be determined in part by the cost *to the firm* of obtaining the necessary knowledge. This is at least in part a function of the firm's own attributes, which can be discussed in terms of the managerial framework.

It is also clear that the discussion of inducement mechanisms in Rosenberg and Moss (see above) bears on our discussion of technological opportunities. The external factors which were seen to be part of the inducement mechanism included opportunities to reduce transactions costs, which were in turn seen to be technologically determined. It is possible to imagine cases in which the integration of two activities such as those explored by Moss might facilitate or require an associated *technical* change in order to accomplish the integration. Such a technical change would be the result of a technological opportunity, whose potential benefits could be expressed in the terms of the Stoneman model above. In such a case the discussion of technological imbalances and opportunities overlap. This underlines a basic theme of the discussion in this section: which is that technological opportunities combine both *incentives* and *constraints*, both of which can operate on the direction of growth of the firm.

2.6 CONCLUSION

This chapter has shown that the discussion of the firm in economics is divided between neoclassical theory which describes the relations between firms but says rather little about their internal components, and managerial theories which say rather a lot about the internal aspects of firms but rather less about the interactions between firms. We have also seen that this general feature of the theory of the firm applies also in some measure to the discussion of technical change and the firm. While the managerial and behavioural approaches seem to us to offer the most potential for further analysis of technical change, in part because they have greater possibilities for internal elaboration and realism, we can agree with Stoneman that the formal analysis based on profit maximising assumptions can also offer important insights.

The initial attempt to apply managerial theories to technical change, reported in this chapter, can be summarised as follows.

Managers have considerable discretion in their choice of strategies to achieve firm growth. Part of the context in which they frame strategies is that of technological incentives and constraints. These incentives and constraints are only in part 'technological', in other respects they are simply economic. Some of these incentives and constraints are derived from the firm's own characteristics which result from its previous history of growth, and some derive from the external technical and economic factors which surround it. Analysis of, and response to, these incentives and constraints may take place in an incremental way with associated internal conflicts in the firm, and in an environment of uncertainty. For the purposes of the continuing analysis in this book we can therefore conclude that managerial decision-making on these topics will be complex, internally differentiated, and therefore susceptible to further analysis. This is the subject matter of Chapters 3 and 4. We can further conclude that the external nature of some of the technological opportunities will imply that they are faced in a similar way by *groups* of firms. This suggests the possibility of some patterns in their behaviour, both with respect to technical change and in general. This possibility has already been raised in Chapter 1 where it was pointed out that a possible level of aggregation in the study of technical change is aggregation with respect to technology type. These topics of patterns in innovation, and in relationships between innovation and industry and market structure are the subject matter of Chapters 5, 6 and 7.

3 Research and Development in the Firm: I. Strategy and Structure

3.1 INTRODUCTION

The previous chapter introduced some issues which arise from the attempt to incorporate innovative behaviour into broader theories of the behaviour of the firm. This chapter and Chapter 4 elaborate some of these issues by discussing in more detail some of the components of innovative behaviour. There are in the literature, a variety of ways of conceptualising the components of innovation and their sequence. Saren (1984) in a review of studies which present 'models' of innovation, identifies six broad classes of model:

1. Departmental stage models.
2. Activity stage models.
3. Linked department/activity models.
4. Decision stage models.
5. Conversion process models.
6. Response models.

Clearly, all of these approaches can be used to present some features of innovation, but they are essentially *ad hoc* in their theoretical character. Our approach does not attempt a new integrative 'theory' of innovation. These two chapters proceed by identifying a number of types or nodes of decision-making, connected with or located in the R & D function within the firm, and concerned with innovation. However, we have not advanced a hypothesis concerning necessary sequences or connections between these different types of decision. To do so would, we believe, impose a rigid structure which does not conform to the reality of innovation. This is not to say that no attempt at theoretical elaboration will be made. On the contrary, for each of the decision types, material is presented which relates the decision to other aspects of the firm's behaviour, and to theoretical accounts such as those outlined in the previous chapter. In some cases these discussions include the suggestion of links between the various types of decisions, but not to the extent of a complete and closed set of linkages. The decision types, or categories of procedure, which are used in these chapters are as follows:

1. The determination of the size of the R & D budget.
2. The division of the R & D budget into major categories of activity and in particular the allocation of resources to basic research.
3. The allocation of R & D resources between broad areas of the firm's activities, for example between divisions.
4. The allocation of resources to general objectives within an area.
5. The allocation of resources to particular projects and products.
6. Procedures for monitoring and control of 1 to 5.
7. The organisation of the R & D function, both internally and in relation to other parts of the firm (including organising inputs from outside the R & D function to decisions of types 1–5 above.

We emphasise again that this is a categorisation for analytical convenience and does not represent a 'decision stage' or 'activity stage' or any other sort of model. Where procedures are linked, we shall comment on those links individually. There is a broad distinction between points 1 to 4 and 5 to 7. The former are mainly concerned with strategic and planning aspects of innovation, whereas the latter are more concerned with organisation and execution. This forms a natural break in the presentation, and thus this chapter will concentrate on the first four topics and Chapter 4 will consider the latter three topics.

3.2 THE THEORETICAL CONTEXT OF R & D STRATEGY FORMULATION

Before looking in turn at topics 1 to 4 above, it is useful to review briefly the sorts of factors that are expected to affect these four types of decision within the received theoretical frameworks available. In this respect it is important to distinguish between the issues and concerns raised in economic theories and those raised in the management literature. These will be summarised and some comparisons presented.

The management literature conventionally presents the corporate planning process as the relevant context for the discussion of all strategic aspects of R & D within the firm. Twiss (1980) for example, argues that R & D strategy should flow from corporate strategy. Bitondo and Frohman (1981) agree and also emphasise the possibility for corporate planning in turn to be influenced by R & D activity. Thus they speak of 'linking' technological and business planning' by means of consideration of 'strategic technology areas'. Merrifield (1981) points out that corporate strategy considerations should influence decisions even at the level of individual R & D projects (point 5 in our list above). One might also argue for the presence of such influences even at the level of organisation and execution of R & D. In view of the pervasive nature of the influence of corporate strategies on all aspects of R & D and innovative behaviour suggested in the management literature, we shall summarise briefly the management approach to corporate planning or strategy formulation. We shall then compare it with economists' approaches to the influences on R & D strategy.

The essence of corporate plans is their role in identifying the markets in which a firm is operating and the markets in which it will be at some point in the future. Related to this is the issue of what product groups it currently produces and how this will change in the future.

The flavour of this approach is well captured by the famous marketing aphorism 'what business are we in?' The range of alternatives then, goes from major changes in the degree of diversification, through to conservative strategies of maintaining an existing (presumably favourable) position in familiar markets. Both strategies may require very specific actions for them to be successful; one does not always remain in the same place by doing the same things.

The choices actually taken are seen as a result of two essentially different types of input. Firstly there must be a 'capability analysis' in which the company forms an assessment of its current strengths and weaknesses. There may be many things the firm would like to do, but some of them may simply be beyond its capacity. It is therefore necessary for the firm to have a clear view of this capacity. For example, a traditional mechanical engineering company may be well aware of the desirability of exploring every possible use for microprocessors in its products, but its total unfamiliarity with such technology will be a real obstacle to adopting this as a full-blown strategy for immediate implementation. Secondly, there must be some estimate of how changes in the environment may alter the spectrum of threats and opportunities which the firm faces. Things which were not possible or necessary in the past may suddenly appear to be so; some of these may fit easily into previous strategies and some may not; some may be achievable with the existing capacity and some may not. Therefore, out of the dialogue between an assessment of the environment and an assessment of the firm's capabilities, a future direction of change can be identified. This may, as has already been pointed out, involve varying degrees of change in the balance between different activities.

These outcomes of the process of corporate planning should, in theory, take the form of a set of specific objectives with specific times by which they are to be achieved. For example, one objective might take the form: 'to remain market leader in product X'; another might take the form: 'to achieve ten per cent market share in product Y by time T'. These objectives might in principle be achieved by a number of routes, but the capability analysis of the company should have enabled it to allocate resources such that particular routes can be specific in advance.

How does this corporate planning approach, which has developed in the management or 'prescriptive' literature (see Chapter 1), compare with the approach to R & D strategy which we might expect to find in the academic economics literature? We have seen in the corporate planning approach, and in the discussion of technological opportunities in Chapter 2, that it is possible to distinguish between sources of influence internal to the firm and sources or influence from the environment. This distinction is important also in the economics literature. However, it is the external influences which have been most systematically analysed in the economics literature.

The dominant paradigm in industrial economics for some years has been the 'structure–conduct–performance' (SCP) paradigm.

(See Scherer, 1980 or other standard industrial economics texts for an exposition.) The essence of this paradigm is to argue that the structure of an industry, that is the number and size of the producers within it, can be characterised on a spectrum which runs from perfect competition through to monopoly. These structural characteristics are then seen as the principal influence on the conduct of the firms in terms of pricing, product differentiation, presence or absence of collusion, and so on, and in turn on the performance of the firms in terms of profits. There are of course many refinements to this basic theme in the literature which are too numerous to mention here: though it is worth noting that many of them involve introducing some reciprocal relationships between the three elements of the model. Thus performance may react upon conduct and structure. The basic direction of causality which is emphasised, however, is *from* external factors in the industry, *to* the conduct of the individual firm. It should be noted, however, that there has been a long-standing debate on the accuracy of one important empirical and theoretical support for the SCP paradigm: namely that concentrated industries enjoy higher profits as a result of collusion. A number of studies have cast doubt on this association (see Reekie, 1979, Chapters 4 and 5.)

But while much industrial economics relates conduct (and therefore by implication R & D strategy) to structural and industrial factors outside the firm, other literature has drawn attention to the particular attributes and specific histories of firms as factors influencing their conduct. A number of studies have attempted to explain why firms facing rather similar market environments respond in different ways. Miles and Snow (1978) for example, distinguish a number of archetypes of strategy which they label 'defenders', 'prospectors', 'analysers' and 'reactors'. As Whitley (1984) points out such classifications are essentially *ad hoc* unless they explain the factors which give rise to such strategic responses. Nevertheless it is clear that they have a theoretical line of descent from the work of Penrose, mentioned in the previous chapter, which emphasises the firm-specific character of the managerial resources available to the firm at any time.

Under the stimulus of the development of the business history literature (Hannah 1976) there has been an increasing attempt to link together the discussion of firm strategy as a resultant of the structural features of its base industry *and* firm-specific attributes.

Chandler (1977) and Caves *et al.* (1980) are substantial examples of this work, and Legraw (1984) has recently extended their framework and added more data. This work is particularly relevant to the analysis of innovation since it takes as its variable the diversification strategies of the firms analysed. While diversification is a much broader concept than innovation, there are many points of overlap in terms of the motivations and influencing factors. Legraw distinguishes four 'strategic groups' in his sample of companies operating in Canada which exhibit different diversification strategies. These are 'single business', 'vertically integrated business', 'related business' and 'unrelated business'. These strategies are then explained *both* in terms of industry variables such as rate of growth, profitability and concentration, *and* in terms of firm attributes, such as strength in R & D, advertising and marketing strengths. His model appears to receive some support from his data, since firms in the sample who followed the diversification strategy 'predicted' by their combination of industry and firm variables achieved higher profits than those which did not follow the predicted strategy.

   In concluding this review of the theoretical context to R & D strategy formulation, we can say that there are some significant points of similarity between the approaches in the management and economics literature to the factors which affect strategy. But whereas the economic analyses are more theoretically grounded, and becoming more susceptible to testing with cross section and historical data, the management literature tends to be more theoretically *ad hoc* and prescriptive. Furthermore, it is clear that progress is made in the discussions in proportion to the attempt to disaggregate various components of strategic behaviour and discuss them in detail. It is with this objective that we now turn to the first four topics on the list of R & D decision levels presented in the introduction to this chapter.

3.3 THE DETERMINATION OF THE R & D BUDGET

This is in many ways the most important and the most difficult of the decisions. The allocation of funds to R & D is an investment decision in that the firm is choosing between present and future returns from its money. The resources allocated to R & D could, in principle, be allocated to the production and marketing of existing products and earn a short-term return. In a firm with a small R & D budget this conflict may not be seen as particularly important, but in very research-intensive firms changes up or down in the R & D budget could involve large sums of money. The fact that many firms do take this step of diverting resources away from potential short-term returns into the longer term activity of R & D suggests that, through a historical learning process, the value of investment in R & D as a longer-term strategic act has become relatively well established in substantial sectors of industry. The criteria for such investment from the economic viewpoint should be the rate of return on that investment. However, the uncertainties of any long term investment, and particularly one which involves the intrinsic unpredictability of some technical events, means that the calculation of a return on the investment is almost impossible *ex ante*, and still open to question *ex post*. This will emerge more clearly from a consideration of the way in which the decisions are typically made in practice.

Taking an example from the management literature, Twiss (1980) suggests the following guidelines through which an R & D budget might be determined:

(a) Costing an agreed programme of research
(b) Comparisons with other firms' budgets
(c) A percentage of turnover
(d) A percentage of profit
(e) Reference to previous budgets.

The first of these, costing an agreed programme, is at first sight the most rational. The elements in a programme of research may have been subject to some form of project evaluation technique (see Chapter 4) and their potential profitability assessed. A well-behaved profit-maximising firm ought to choose to do only those projects which yield a return greater than that which can be achieved by investing elsewhere, and set its total R & D budget as the sum of the costs of those projects. In practice this is impossible. Firstly, the project evaluation techniques do not provide information with that degree of reliability. Secondly, even if such techniques are not used, the number of projects which the R & D staff would like to do will often exceed the amount of finance available, simply because many projects are inherently attractive on purely technical grounds, but their commercial viability is still unknown. This type of 'bottom-up' approach to budgeting is therefore impractical, and some sort of aggregate 'top-down' approach is more common. The significance of 'top-down' budgeting is discussed further below.

The second technique, comparisons with other firms, has certain attractions. It seems that many industries have established over the years fairly stable average levels of research intensity. This suggests that in a given industry with a given technological base (and perhaps at a given stage of development) there is an 'appropriate' or minimum level of R & D which is almost a 'natural' requirement. Thus, observation of other firms may be a useful way of determining whether a firm is spending too much or too little by a vast amount. Beyond this general point however, its limitations as a technique are considerable. The information on other firms' budgets is not always available or easy to interpret. Definitions of R & D may vary from firm to firm, different types of programme have different intrinsic costs, and different firms may be more efficient in their use of R & D funds. Furthermore, the differing degrees of diversification between firms means that there is rarely an absolutely perfect firm to use as a yardstick. Overall however, there can be no doubt that observations of other firms can play some part in judging an appropriate level of R & D.

Relating the R & D budget to the turnover is an easy-to-use formula which appears to be quite common, not least because it is the usual way of measuring research intensity. In purely logical terms it suffers from a grave fault in that it relates an expenditure to a revenue which is the result of some previous expenditure, rather than to a future revenue. In its favour is the fact, which may be rational, that as turnover grows, the R & D budget will grow in line with the size of the company. A similar technique is to set the R & D budget at some percentage of the profit of the company. The problem with this technique is that profit can fluctuate much more than turnover on a year to year basis, and this would imply a considerable instability in the R & D budget. However, expensive highly-trained personnel and equipment working on long-term projects cannot be simply switched on and off like factory lights. Nevertheless, it is clear that profit has some sort of effect on levels of R & D since prolonged profit squeezes tend to reduce overall R & D activity and to shorten its time-horizon. This was particularly true in the UK during the 1970s (Bosworth, 1979; OECD, 1980).

Finally, there is the 'institutional response' which determines a budget by first and foremost looking at the previous budget and arriving at an incremental change through a process of negotiation. In this process the R & D director is likely to try to argue the budget up on the grounds of inflation, the need for new equipment and personnel, and promising new projects. The controllers of the resources, whatever their level of objective agreement with the arguments of the R & D director, will have an inbuilt tendency to whittle the proposed increase down because they need to balance the claim against competing claims.

Considering these alternative methods for setting the R & D budget, it becomes clear that they are not really alternatives at all. They each have a certain plausibility because each refers to one part of the overall set of constraints and incentives which contribute to the process of setting the budget. To ignore what other firms are doing, or the availability of retained earnings, or the actual programmes already in progress which structure activity, would be absurd. Equally it would be absurd to determine the budget by reference to only one of these factors. It seems then, that we are forced to conclude that the final R & D budget figure emerges as a result of the complex interaction between these different methods of judgement. But we have also noted that the levels of R & D tend to exhibit some uniformity at the industry level, which implies that the budget level is not determined by forces unique to each firm. Thus we return to the familiar problem of theoretically combining the internal and external influences. One interesting approach to this task takes the observed *stability* of R & D budgets over time as the phenomenon to explain, and utilises modified behavioural and managerial theories of the firm. This is the approach of Kay (1979), who begins by examining the applicability of Cyert and March's theory. They see innovation, and by implication allocations to R & D, as falling under the two headings 'distress' and 'slack'. The former is innovation serving organisational goals in response to a specific problem. The latter is innovation serving sub-unit goals and is in a sense 'pure' organisational slack (Kay, 1979). But this model, according to Kay, should generate fluctuating and unstable R & D budgets because the incidence of the problems which trigger the distress component in

R & D is unpredictable. However, Cyert and March themselves see R & D as a sub-unit with a great capacity to achieve smoothing of budget allocations from year to year, suggesting the operation of one of the decision rules or standard operating procedures which form an important part of their theory.

Kay sees this as a contradiction in the Cyert and March theory which indicates a deficiency. He develops a modified theory based on the systems approach of von Bertalauffy (1973) in which he sees the distress innovation of Cyert and March as short term stimulus–response behaviour not characteristic of the higher level in the systems hierarchy which he associates with the senior management of a company. While such stimulus response behaviour may exist in the short term in the operating units of a company, Kay sees differently the long term strategic behaviour at the top level of management. He argues that the senior managers have a stable preference system which is linked to a way of seeing the company which Kay likens to a 'gestalt' or pattern. This framework of preferences is the result of the specific attributes of the management and the firm and their experience both as individuals and as a company, and creates a powerful organising influence on planning decisions. One of its outcomes can therefore be seen as a stabilising influence on the level of R & D in a manner which is essentially 'top down' in character (see above).

The question remains however, what determines the actual level around which the stable preferences stabilise? Kay presents some empirical analysis of the R & D budgets of a sample of firms and finds three independent variables which are positively correlated with R & D budget size. These are the past growth rate of the firm, the size of the firm, and the technological opportunity of the industry in which the firm is operating. The last variable is particularly interesting, since it gives further support to the analysis of technological opportunity in the last chapter. It should be added though, that there are great difficulties in measuring opportunity as an independent variable. Nevertheless this seems to be an encouraging example of the benefits of using a perspective in which industry and firm-specific variables enter into a model predicting an aspect of firm strategy. It is particularly interesting that, although firm size enters the model, industry structure does not. The main industry variable is technological opportunity, which strengthens the case for recognising aggregation by technology as a useful method of analysis in some circumstances, as was pointed out in Chapter 1.

A model which does combine technological and structural characteristics of the industry is developed by Stoneman (1983). In this model R & D expenditure is allowed to generate cost reductions (process innovation), or increases in demand (product innovations), and the latter can be in part determined by expectations concerning the behaviour of rivals. The conditions for profit maximisation of the model present some interesting results. R & D as a percentage of sales is found to be positively related to technological opportunity (which enters the model as the slope of the cost-reducing and demand-generating functions of R & D), and negatively related to the expected degree of retaliation by rivals.

Stoneman does not test the model, but does review other empirical literature which has tested the relationships between R & D and technological opportunity and between R & D and market structure. While the former receives some limited support the latter presents a more complex picture. Detailed discussion of the innovation/market structure relationship is deferred to Chapter 5 of this book. For the present discussion it is noted simply that the balance of opinion sees technical characteristics of the industry as an important factor influencing R & D intensity, but finds difficulty in empirically separating its effect from the possible effects of structure. Stoneman does not however, discuss the work of Kay, mentioned above, which does suggest an independent confirmation of the role of technological opportunity.

The preceding discussion has suggested some promising lines of development in research on the role of R & D expenditure in firm strategy. Further research is clearly needed in the elaboration of the concept of technological opportunity, and in empirical studies, particularly of a longitudinal character.

3.4   THE ALLOCATION OF RESOURCES TO MAJOR ACTIVITIES WITHIN R & D

It has been noted that the allocation of resources to innovation is in itself a movement into a realm which is more uncertain than the conduct of the existing lines of business. Within the sphere of innovative activities themselves there is a range of degrees of uncertainty, and this influences the allocation of resources to these different types of innovative activity. (Freeman, 1982).

The activity which exhibits greatest uncertainty and creates the most difficulty in decision-taking is basic or fundamental research. There can be no doubt that without basic research, which advances the spectrum of scientific and technical possibilities, there can be little innovation in the long run. From the point of view of the individual firm, however, basic research has some fundamentally unattractive qualities. First, it takes a long time and its outcome is more uncertain than any other R & D activity. Secondly, the results of basic research are not so easily contained within the firm in the same way that inventions can be contained, since pure scientific knowledge cannot be patented. These peculiarities of basic research have led to the view that it should be undertaken principally by the universities and the state, as a compensation for what would otherwise be an underinvestment in science at the level of the nation-state. (Norris and Vaizey, 1973). It is undoubtedly true that this is one real reason for the development of state-funded science (see Chapter 8). It is important to note however, that firms cannot successfully use the public results of state-funded science if they do not themselves have some pure science capability. A simple example will illustrate this point; in order to scan effectively the journals in which the results of basic science are published, one needs some of the competences of the field itself. These can only be sustained if the individual concerned is a participant in the field. This is not to suggest that all firms doing R & D need also to do basic research; in some industries the relevant technologies are further removed from genuine scientific frontiers than others. Nevertheless it is clear that basic research can have a role as a transmission channel for information into the company, which broadens the pool from which it gains ideas in a very significant way (Gibbons and Johnston, 1974).

In addition, when basic research produces results which are usable (following whatever applied research and development work is necessary to apply the idea) it often – though not always – produces a competitive lead or quantum jump in product character which is much more fundamental than those which result from more routine innovation. We are touching here on the large and difficult issue of the traditional distinction between radical and incremental innovations. Basic research is not always necessary to radical innovation, but it is often associated with it.

Kay's model of R & D resource allocation, discussed above, gives further insight into the issue of spending on basic research. He considers many of the same disincentives to the performance of basic research which have been noted above, and concludes that it will be the lowest priority activity amongst sub-areas of R & D within the managerial preference system. Thus he predicts that when the total R & D funds are subject to some constraint such as a shortage of retained earnings, then basic research will be cut before applied research or development. Kay's own data do not give clear support to this prediction, but the work of Bosworth (1979) and Freeman, Clark and Soete (1982, Ch. 4) does support the notion that severe long-term recessions depress R & D and basic research, but that short-term recessions do not. (See Chapter 7 below for further discussion of long-term recessions).

3.5 ALLOCATION OF RESOURCES TO BROAD AREAS

The allocation of research resources discussed so far have been largely seen in terms of likely long and short term returns and the problem of balancing between them. (This point also applies to the discussion of individual projects in the next chapter.) There is another dimension along which choices have to be made in allocating R & D resources, which could be described as a dimension of diversity. Except for the increasingly rare case of the one-product company, most firms are engaged in a number of fields, and they will be aware that this number could change in the future.

We have noted in Chapter 2 that such firms may opt to organise in the M-form rather than the U-form in order to benefit from placing different groups of products in separate divisions. In such cases the question arises whether to divisionalise the R & D function or to retain it as a central function. Kay (1979) assumes that the latter is typical, but in fact both solutions are observed. In later work Kay has re-examined this point, and further discussion is deferred to the next chapter where organisational issues are treated in more detail.

Wherever the R & D function is located in the firm, it seems clear from the preceding discussions of total budget that senior management will retain a significant involvement in allocations between major areas. Thus while at the extreme, divisions of some companies may have complete autonomy in the determination of their R & D policy, this is not likely to be the general case. It is in the divisions that we would expect to find the different emphasis on growth, or consolidation, or cost cutting that characterise the detailed components of a corporate plan. If there are specific objectives for each division, then this ought to have some explicit implications for the use of R & D resources.

Thus the assessment of technological opportunities and capabilities which characterises the R & D budget at the aggregate level applies in a different scale at the level of the product group or division. The balance between these competing claims becomes difficult because there are no universal standards of comparison. It is rather difficult to compare scientifically the commercial prospects of two entirely different types of R & D work in unrelated fields, but some comparison has to be made in order to set budgets.

An especially difficult case of this type of decision is that which involves entering a completely new product area where existing divisional structures are inappropriate in the functions of production and marketing as well as R & D. Often this step might be taken by the expedient of acquiring a whole company, thus avoiding some aspects of the problem, but this is not always possible. An example will help to illustrate this point. Most observers agree that one of the fastest growing markets in capital goods over the next twenty years will be for office automation technologies. There is already a significant number of firms in the field but the technology is not yet completely stabilised and there is still scope for large firms to enter the market. The range of firms who could consider entering runs from computer firms, through office equipment firms not yet making computer equipment, to general electrical engineering companies. If they do make a step in this direction they need to consider a number of extremely difficult issues such as what the ultimate shape of that market might be, and whether they will need to diversify further into data transmission technologies in order to remain in the market, and whether these further steps also fit into their broader objectives. In the case of a giant company such as IBM this direction of diversification may even take it away from its traditional base, the mainframe computer (*Economist*, 4 May 1985).

Economic analysis of the allocation of R & D resources between areas and divisions remains under-developed. In principle, it should be possible to apply the models of total R & D budgeting developed by Kay and Stoneman, and discussed above, to the area or division level, if technological opportunities can be identified and specified at the same level of disaggregation. There appears to be no work in this direction however. Legraw's (1984) analysis of diversification strategies points out that in the case of 'related business' diversification, the desire to use firm-specific surplus R & D resources may be an important incentive. This could be a useful insight to test in case study work.

A rather different argument concerning allocation to R & D areas is that which emphasises the potentially protective character of R & D for the dominant or monopolist firm. R & D can create an entry barrier to an industry not only when embodied in existing products, but also when 'kept on the shelf' as potential products to deter rivals. There is considerable anecdotal evidence of this practice, and Cowling (1982) argues that it is a consistent feature of the investment strategy of oligopolists. However, there is no conclusive evidence on whether the products 'on the shelf' are the result of purposive R & D or the by-products of R & D with other primary objectives. In practice these things are difficult to separate even in the most careful case studies.

What emerges clearly from this discussion of allocation to R & D areas is that it interacts strongly with the other decision types in the list introduced at the beginning of this chapter. In particular there is clearly some mutual determination between this decision and the setting of the total budget. In addition, it seems reasonable to expect allocation to areas to be related to the nature of the objectives being pursued in different product groups and technologies. This is the next and fourth level in the list, and we now turn to examine it in detail.

3.6 ALLOCATION OF RESOURCES TO OBJECTIVES WITHIN AREAS

Within a particular product group or division, R & D does not proceed by a random process of unconnected individual projects.

It is more common to find that groups of projects relate to broader objectives which can not be achieved by one route alone. For example, the objective of fuel efficiency in motor cars has become much more important in recent years. It can be achieved by a variety of different changes in engine design, in drag coefficient, in weight of components and so on. These different approaches call on an enormous range of different types of R & D activity. But there will also be projects and resources geared to other general objectives such as safety, or more general notions such as consumer appeal. These objectives therefore play a part in determining the distribution of R & D projects at the disaggregated level, and they also play a part in the assessment of the fundamental possibilities and demands of different broad areas and product groups, as was discussed under 3.5 above.

The factors which give rise to these general objectives, and to changes in their perceived ranking is perhaps one of the most interesting areas in the study of innovation. Clearly, the change in oil prices was a major influence on the rising importance of fuel efficiency in cars. But the specifically political factor of the environmental lobby also played a significant part in changing R & D priorities in the car industry, (Johnson and Gummett, 1979). The balance between scientific factors and demand-side influences, and their relation to social and political influences has long been discussed in the innovation literature (these arguments are reviewed in Chapters 5, 8, 9 and 10). For the moment we shall concentrate on the phenomenon of a number of similar innovations which appear to result from the enduring nature of certain objectives perceived within the firm.

Once an idea like fuel efficiency becomes established as an objective, it develops almost a life of its own. It acts as a point of competitive comparison between companies, and it acts as a day-to-day paradigm to orientate the way engineers performing R & D see their task. It may even create a climate in which this is seen as a more important strategy for 'improving' the car than other areas of improvement which could be considered.

The relationships between the performance factor 'fuel efficiency' and the contributing specification parameters such as 'drag coefficient', 'rolling resistance', 'combustion efficiency', 'power to weight ratio' and so on, might therefore be seen as a series of linked 'natural trajectories', or intensive technological opportunities such as those described in the last chapter. In each case there are certain well-understood technical relationships which govern the extent of possible improvement; familiar knowledge and skills which may be used to extract those improvements; and in some cases, the application of less familiar knowledge to make other improvements which may be more uncertain and more costly. Over a period of time the 'Penrose effect' could be expected to occur to some extent in these circumstances, resulting in some specialisation of the attributes of the firm's R & D function around these natural trajectories.

The fact that the ultimate performance variable fuel efficiency is, in this case, the resultant of the evolution of several contributing natural trajectories illustrates the possible occurrence of the imbalances or focusing effects discussed by Rosenberg and Moss. As one avenue of technical improvement yields some gains, the incentives surrounding the allocation of effort to another trajectory may change. Thus if drag coefficients cannot be reduced much below 0.3, then more effort may be put into weight reduction or some other area of change. The shifting location of the bottleneck is a feature of the fact that the product is a complex system of technical relationships which interact with each other.

It is clear that the presence of this tendency to group R & D projects together under the headings of broad objectives is a powerful organising influence in the conduct of R & D. It creates continuity from year to year in at least some of the planned activity, thus creating some base line in the budgeting process, some sources of demand for the activities in the basic research sector, if such exists, and some momentum in the development of the R & D activity of that division or product group. Of course this source of structuring of R & D is only one of the influences. New factors, changes in the perceived importance of existing factors, and other disturbances will play a role in changing the pattern. It is important to recognise however, that there are some relatively enduring factors such as those discussed in this section. Any economic analysis of innovation must take these into account. Their influence on other aspects of R & D behaviour will be clear from consideration of their interactions with the other decision types in the list presented at the beginning of this chapter. The contribution of such behaviour to the interactions between firms is deferred to part II of the book.

3.7 CONCLUSION

This chapter has considered four aspects of decision-making in R & D which related to the strategic acts of firms. These four aspects are: the level of R & D, the degree of basic research, the structure of the R & D programme at the broad level, and the range of objectives pursued in the programme. Some theoretical context was considered in the form of the structure–conduct–performance paradigm of industrial economics, the business history approach to firm development, the received wisdom on the corporate planning process, and the discussion of technological opportunities from the preceding chapter.

The discussion of the four types of decision making has given some support to the view that R & D strategies can usefully be described in these terms; with properties of size, degree of radicalness, structure, and direction. The four properties are mutually determining to some degree, because they are each in part influenced by some of the same factors. These factors include firm specific resources, technological opportunities and industry and market structure. Although there is some 'self-generating' component to strategy resulting from the momentum of previous allocations, the predominant location of strategy formulation seems to be at the higher levels of management and to be the result of qualitative perceptions of an appropriate 'pattern' for the company, rather than purely financial assessments of discrete projects. Despite this behavioural account of the mechanisms which generate the R & D allocations, it seems that there are promising possibilities for the development of models which can predict some of the results of this behaviour. This work is so far mainly limited to the level of R & D allocation, but extension to the structure and direction of R & D is a high priority area for further research.

Throughout this chapter we have consciously avoided referring to the typologies of innovation strategies developed by some authors as attempts to describe certain regularities in the observed variety of innovative behaviour. This is not a reflection on their usefulness but rather a result of a desire to start from some components of strategy and then progress to broader generalisations. The most useful, empirically elaborated, and often quoted typology of innovation strategies is that of Freeman (1982, Ch. 8). He defines offensive, defensive, imitative, dependent, traditional and opportunist strategies, relating them to character of R & D, phase in product cycle, balance of R & D with marketing or production strengths, and a number of other factors. As he acknowledges, his typology is not meant to act as a theory of the firm, nor is it derived from that type of analysis, but rather from the considerable case study literature on innovation which permits some degree of inductive generalisation. What relationship exists between this typology and the different style of analysis presented in this chapter?

In considering this question it is important to make one proviso. An innovation strategy and an R & D strategy are not the same thing. Freeman's discussion includes all aspects of the behaviour of a firm which may be part of its innovative behaviour. For example, the importance of marketing actions which are consistent with the innovation strategy has been identified in the SAPPHO project (1972) and this point is incorporated in Freeman's discussion of defensive strategies in particular. The discussion in this chapter has concentrated on those aspects of innovative behaviour which relate more directly to the R & D programme, though clearly there are many points of contact with the other dimensions of firm behaviour. Notwithstanding this difference in inclusiveness, (which will be narrowed by the discussion of organisational considerations in the following chapter), we can see many similarities between the issues discussed in this chapter and those which go into the formation of Freeman's strategy types.

Freeman regards his strategies not as characterising a complete firm in all its lines of business at all times, but rather as descriptions of partial activities, limited both in time and scope. Thus, in the terms of the discussion above, it is possible to see them as the result of the operation of the various factors we have noted. High technological opportunities, relevant R & D expertise and related managerial expertise, and low entry barriers, would seem to be good predictors of offensive strategy in a particular area of a company's business. If these characteristics related to a process technology rather than to a product technology however, it might be more indicative of a defensive strategy. Such combinations of factors and labelling as strategic types is an activity which could usefully guide future case study work in innovation.

But does strategy in the Freeman sense exist only as the *result* of the factors discussed above, or is it something extra, perhaps a '*residual*' or unexplained aspect of the firm's behaviour which is not directly attributable to the explanatory factors, whether internal or external, in the models developed in the literature? There is perhaps some space in the explanations of strategy for the notion of 'culture' in a firm. Some firms are often described by researchers and business journalists as having a distinctive culture which is unlike that of other firms and is a product of deep-rooted influences from its foundation, founders and attitudes to the practice of business in the broadest possible sense. While these attributes could theoretically be included in the category of 'firm-specific managerial resources', it may be more useful and more interesting to explore the notion of firm culture explicitly, both as an extra influence on innovative and diversifying behaviour, and for its own sake. Such studies however, are probably more suited to the sociological study of management than the economic analysis of innovation.

4 Research and Development in the Firm: II. Organisation and Execution

4.1 INTRODUCTION

This chapter is concerned with the remaining three topics on the list of decision types introduced at the beginning of the preceding chapter. These are: the allocation of resources to R & D projects, the monitoring and control of projects, and the organisation of the R & D function, both internally and in relation to the other parts of the firm. These issues are therefore more closely related to the operational aspects of R & D than those discussed in previous chapters. This operational character has resulted in rather more attention being given to these topics by the management literature and rather less by the academic economics literature. However, it is through operational activities that strategic intent is expressed, if a rationalist view is taken; or is revealed, if a somewhat less rationalist view is taken. Therefore the connections between operational issues and the other decision types discussed previously should always be borne in mind.

4.2 R & D PROJECTS

Origins

R & D projects are the basic unit in the conduct of R & D. By project we mean a piece of scientific or technical work which has a specific objective or area of enquiry, a budget, and an expected duration, and is the responsibility of named individuals. In reality the degree of formality with which these requirements are fulfilled will depend on the nature of the firm, the size and degree of differentiation of the R & D department, the preferences and management styles of the people in immediate authority over the department and, to some extent, on the nature of the project itself. Thus basic research projects, where they exist, may be less closely defined in all respects than development projects. Such a difference would reflect the fundamentally different levels of uncertainty which characterise the activities of scientific research and technical development. The issue of uncertainty is further discussed below.

While the bulk of the activity in an R & D department will be organised into projects, there may be some activities which are less organised. For example, some firms may choose to give some of their scientific staff some time to pursue research which is self-directed. The motivation for this may be to allow scientists to maintain contact with a field or discipline by similar means to those used by an academic scientist, or it may be to give the scientist an incentive to use the time to initiate new projects of direct commercial relevance to the firm. Once again, the occurence of such practices is very variable, and dependent on the nature of the area of science or technology being investigated.

The role of basic research has already been considered briefly in the preceding chapter. In this section we shall concentrate on those R & D projects which have an applied research or development character, as being the most relevant to considerations of operational management in R & D. This is also the type of project which is most discussed in the literature.

From the point of view of the student of technical change, the identification of the origins of R & D projects is a task which overlaps considerably with the identification of the origins of inventions and innovations themselves. However, the two tasks are not the same. An innovation may consist of the results of a great many discrete R & D projects which contribute to different parts of the innovation. Successful innovations may result from a group of R & D projects of which many were unsuccessful, and successful R & D projects may contribute results to innovations that are ultimately commercially unsuccessful. It may even be that projects initiated with one innovation in mind are eventually used in an entirely different innovation. Nevertheless, the overlap between the two topics is substantial.

In the case of innovations themselves, discussion of origins has traditionally been conducted in terms of the two influences of 'science push' and 'demand pull'. The former approach is closely identified with the Schumpeterian tradition in research on technical change, while the latter is found both in the economic literature (Schmookler, 1966) and in the case study literature. However, recent research has suggested a number of non-trivial ways of analytically combining the two influences, with due allowance for specificity of technical and commercial circumstances. This work is reviewed in Chapter 5 of this book. In the case of R & D projects the presence of scientific and commercial factors in the decision to initiate a project is no less important and no less complex. However, as will be clear from the preceding discussion, the level of aggregation of the problem is not the same since R & D projects are, in general, on a smaller scale than innovations, and by definition are conducted inside the firm rather than at the interface of the firm and the market.

Recalling the simple model of the innovating firm presented in Chapter 1, we can expect the sources of stimuli to R & D projects to be distributed across the other principal functions of the firm such as marketing and production, as well as from within the R & D department itself. Some examples of these sources can now be presented briefly, followed by an attempt at classification.

The first case which can be considered is one in which the initiative comes from a perceived problem in an existing product or process. This could range from a minor difficulty such as a part which fails earlier than its designed life-span, through to a major difficulty which completely nullifies the function of the product. Although the personnel who first identify the problem could in principle be in any part of the company, it is more likely that they would be in marketing (in the case of a product) or production (in the case of a process). If the product is itself in its pre-production development phase then it may be R & D staff who first encounter the problem. This type of 'de-bugging' activity generates a very large percentage of R & D projects, though the proportion will obviously vary from industry to industry.

A second common source of project generation is a decision, often contingent on a marketing judgement, to change a particular specification of a product by a modest increment. This is particularly common in products with multiple characteristics sold in markets characterised by non-price competition and product differentiation. Examples of this source of project generation abound in the motor car and other consumer durable industries, where an extra increment of power, of fuel efficiency or of signal-to-noise ratio are frequently observed improvements. There is clearly a connection between this type of project origin and the discussion of natural trajectories of technical change in the preceding chapter. Whereas the idea of trajectory was related more to the objective possibilities of the technology, or the specificity of R & D resources, the example above introduces the possibility of choice *between* trajectories being influenced to some extent by marketing judgements.

The type of project mentioned above is subject to considerable variety in its level of 'radicalness' with respect to the existing technology of the firm and the industry. The more radical the project, the more it overlaps with a third type of project which is that concerned with a new product. R & D projects directed at the creation of a new product are likely to be related to more fundamental judgements concerning the suitability of the existing product range, the need for further diversification or integration, and other strategic considerations. The scale of such projects usually results in their rapidly being sub-divided into a family of component projects with a consequent increase in the division of labour among R & D staff and multiplication of management problems in monitoring and control.

The three broad types of project generation mentioned above are not an exhaustive classification but an illustration of the dimensions of differentiation. Attempts at classification are hazardous because they tend to over-simplify the tremendous variety and individual character of all changes in technology. Nevertheless, it is useful to list those dimensions along which the factors which stimulate R & D projects can be differentiated.

1. Internal or external to the firm.
2. Initiative through marketing, R & D, or production staff.
3. Use of an existing technology in a new way, or development of a new technology, or improvement of a technology.
4. Based on a realisation of a technical opportunity or on the need to solve a pressing problem.
5. Resulting from development of existing product range or from a desire to change product range.
6. Active or reactive decision.

Project generation is therefore a very diversified set of processes. It has an active and a passive component, it is fed by stimuli from many sources and produces project ideas of widely differing levels of importance.

Project evaluation and selection

The number and diversity of the ideas for R & D projects in any but the smallest or most specialised company will necessitate the use of some procedure for the evaluation of these ideas and the choice of those which are to be undertaken. A great proportion of the ideas may never be taken up at all since there will normally be some preliminary screening process which weeds out those ideas which are not deemed to be worthy even of further evaluation. Screening may be more or less formally organised, ranging from the personal decision of an R & D manager, through to formal consideration of short written proposals by a committee of managers. Project ideas which survive screening, or which flow more directly from the explicit strategy and objectives of the R & D department should, according to the management literature (see for example, Twiss, 1980, ch. 4) be submitted to some formal evaluation procedure. We now consider the nature of such evaluation procedures and the difficulties they encounter.

R & D projects are investment projects which are fundamental to the future of the organisation. Therefore the type of evaluation which is in principle most appropriate is an investment appraisal. This entails the need to assemble estimates of expenditures, revenues and timings. As has been discussed in Chapter 3, however, it is also the case that R & D projects contain the seeds of a *direction* of growth for a company as well as a *level* of growth. Therefore they need to be appraised and ultimately selected on the basis of qualitative criteria concerning the strategy of the company as well as on conventional criteria of return on investment. The principal obstacle to the assembly of both the quantitative and qualitative data for the evaluation of projects is the inherent uncertainty of R & D in particular and of innovation more generally.

Since Knight (1921) it has been conventional to distinguish between uncertainty and risk. Risk is an essentially statistical concept which is related to the distribution of a population of homogeneous events. Thus if all R & D projects and innovations were characterised by similar rules relating inputs to outputs, then risk factors could be calculated and suitable 'insurance' provided against the likelihood of failure. However, R & D projects are not homogeneous events, and while there may be some possibility for learning from experience, particularly in the field of development work, the bulk of R & D is characterised by 'true' uncertainty which cannot be insured against.

Freeman (1982) has suggested that the uncertainty in technological innovation consists of a technical component, which concerns the unpredictability of the timing and character of scientific and technical events; and a market component which concerns the complex changes in the structure of demand. Added to these is the non-specific but equally important element of 'general business uncertainty', which refers to the broad economic climate, and which is usually handled by investors by the application of a 'suitable' discount rate to all future expenditures and revenues.

Freeman also follows many other writers in pointing out that not all innovations or R & D projects are subject to the same degree of uncertainty. There may in principle be a spectrum of uncertainty stretching from basic research at the highest level, through radical product innovations, new versions of existing products, new processes, licensed innovations, product differentiation, to minor technical improvements. Such a scale however, while acceptable as a generalisation, conceals important variations. For example, a minor improvement in a product's performance may be dependent on a more radical change in one of the technologies of the subsystem which is responsible for the improved performance. The use of the terms 'radical' and 'incremental' innovation and 'more or less uncertain' is very dependent on the definition of *to whom* the uncertainty is relevant. The use of a micro electronic control device to substitute for a conventional electronic control circuit for example, may well be quite low in uncertainty to the innovator of the product, while the development of the circuit itself may have been much more radical to the circuit producer.

Furthermore, although it is true on average that basic research is more uncertain than applied research, and this more uncertain than development work, it is a familiar feature of R & D projects that the focus of attention, or bottleneck, in the project may shift back and forth in time between research and development as the solution of one problem creates other problems at different levels.

The role of uncertainty has been emphasised because of its endemic character in innovation. In the light of uncertainty it is perhaps even clearer that R & D budgeting will tend to operate in a top-down manner, as was argued in the previous chapter, rather than on the basis of aggregating the budgets of large numbers of inherently uncertain projects. Nevertheless, uncertain though they are, projects have to be financially evaluated and given budgets, if only for the purpose of control. We therefore turn now to some of the techniques used for evaluation.

Many authors have proposed techniques for the evaluation of R & D projects which often involve formulae which have to be applied to various parameters of a project and a resultant merit number calculated. A representative example is that of Hart (1966):

$$Project\ index = \frac{S \times P \times p \times t}{100\ C}$$

where S = peak sales volume £ p. a.
P = net profit on sales (%)
p = probability of R & D success $(0 - 1)$
t = a discount factor
C = future cost of R & D.

The formulae of other authors differ in detail but not in basic approach, and a review can be found in Baker (1974). More recently, some authors have partitioned their formulae to deal with different stages of the life of a project (Muncaster, 1981), including when to terminate projects (Balachandra and Radin, 1980), and in order to cater for projects with different levels of development or research activity (Becker, 1980). Such developments represent the attempt to incorporate experience more explicitly into the management procedures recommended in R & D, though there is little evidence that this is having any effect, positive or negative, on the efficiency of project evaluation.

The common feature of these evaluation procedures is that they all attempt to capture two features of a project. These are its likely cost-benefit ratio and its probability of success. It is interesting to note that a high score on one factor may often, though not always, imply a low score on the other. Both of these factors can only be estimated by the assembly of a large volume of quantitative and qualitative data about the project. The calculation of a final merit number is simply a way of putting a great many essentially incommensurable parameters into one number, to facilitate consideration of projects one against the other. Clearly it is the data itself which is of the most interest, since the weights in which they are combined are essentially arbitrary. There is therefore reason to believe that the real significance of evaluation techniques is the discipline which they may impose on data collection, rather than the presentation of the data in summary form as a merit number. The role of evaluation and selection techniques as a discursive process will be further discussed below. Firstly, it is appropriate to consider briefly the nature of the quantitative and qualitative data which may be required for the evaluation process.

The first element in the quantitative data is an estimate of the costs and duration of the research and development work. The difficulty of this task will depend on the degree to which the work is familiar to the firm itself, or to the field in general. As noted above, whilst all technical investigation and development is intrinsically uncertain, it may be less uncertain if the problems, techniques, equipment and forms of solution are part of a class of problems which have been attempted before in the firm. Clearly related is the extent to which the experience is embodied in currently available resources within the firm.

Take for example, the applied research to produce a modified version of an important drug, where the objectives might be to improve efficacy, reduce side-effects, and improve production economies. This research would involve some very specific chains of events which can, to the experienced pharmaceutical R & D scientist, *to some extent* be planned out in advance and therefore estimated. A range of scientists, technicians and other staff will be employed to produce modifications to existing drugs, to synthesise and purify them, to test them on animals, to design and implement clinical trials, and so on. By looking at past experience and at the specific features of the new project a total cost for applied research and development could be estimated. This would consist of increments spent over a series of budgeting periods into the future, and therefore it would be appropriate to reduce these to a Net Present Value by the application of a discount rate. At a later stage, investment in new production equipment may be needed, as will expenditure on marketing. These nevertheless have to be estimated at the initiation of the project, even though they may be intrinsically more difficult to foresee because of their greater distance in time.

Secondly, there is obviously a need to consider the market. This involves considering the size of the potential market, its rate of development, potential competition, likely market share, scope for segmentation and product differentiation and similar factors. A time profile of the development of sales and an estimate of the initial and subsequent price structures are what the estimator ideally needs. Much of this quantitative data is deduced, however, from qualitative data. Some authors have concluded from this that it is this type of data which is the most essential in assessing projects, and have therefore attempted to include as many such items as possible in 'check lists' for the evaluation of R & D projects. These can be variously seen as alternatives or complements to the formulae discussed above. Examples can be found in Twiss (1980) and Freeman (1982). The example from Twiss is reproduced as Table 4.1 to illustrate the variety of factors which may be included, and to act as a basis for a brief discussion.

TABLE 4.1   *Check-list of project evaluation criteria*

A.   *Corporate objectives, strategy, policies, and values*
1.   Is it compatible with the company's current strategy and long range plan?
2.   Is its potential such that a change in the current strategy is warranted?
3.   Is it consistent with the company 'image'?
4.   Is it consistent with the corporate attitude to risk?
5.   Is it consistent with the corporate attitude to innovation?
6.   Does it meet the corporate needs for time-gearing?

B.   *Marketing criteria*
1.   Does it meet a clearly defined market need?
2.   Estimated total market size.
3.   Estimated market share.
4.   Estimated product life.
5.   Probability of commercial success.
6.   Likely sales volume (based on items 2 to 5).
7.   Time scale and relationship to the market plan.
8.   Effect upon current products.
9.   Pricing and customer acceptance.

10. Competitive position.
11. Compatibility with existing distribution channels.
12. Estimated launching costs.
C. *Research and development criteria*
 1. Is it consistent with the company's R & D strategy?
 2. Does its potential warrant a change to the R & D strategy?
 3. Probability of technical success.
 4. Development cost and time.
 5. Patent position.
 6. Availability of R & D resources.
 7. Possible future developments of the product and future applications of the new technology generated.
 8. Effect upon other projects.
D. *Financial criteria*
 1. Research and development cost:
    (*a*) capital.
    (*b*) revenue.
 2. Manufacturing investment.
 3. Marketing investment.
 4. Availability of finance related to time scale.
 5. Effect upon other projects requiring finance.
 6. Time to break-even and maximum negative cash flow.
 7. Potential annual benefit and time scale.
 8. Expected profit margin.
 9. Does it meet the company's investment criteria?
E. *Production criteria*
 1. New processes involved.
 2. Availability of manufacturing personnel – numbers and skills.
 3. Compatibility with existing capability.
 4. Cost and availability of raw material.
 5. Cost of manufacture.
 6. Requirements for additional facilities.
 7. Manufacturing safety.
 8. Value added in production.
F. *Environmental and ecological criteria*
 1. Possible hazards – product and production process.
 2. Sensitivity to public opinion.
 3. Current and projected legislation.
 4. Effect upon employment.

Note: This check-list is not comprehensive and suitable for universal application. Nevertheless most companies are likely to find the items listed relevant in project evaluation.
Source: Twiss (1980)

In section A of Twiss's list many of the issues considered in the strategic categories of decision discussed in the last chapter are raised. For example, questions of risk and image (3 and 4) are related to the basic strategy and diversification plans of the company. Questions of time-gearing relate to the achievement of long term performance objectives within areas and to issues of product obsolescence in existing markets. In short these issues are of the fundamental type: 'Is this project pointing in the direction we regard as appropriate, regardless of its supposed cost/benefit ratio?'

Sections B and C of the check list deal mainly with issues related to cost/benefit ratios, but also with issues of capability and feasibility (for example C6; 'Do we have the R & D skills?', C8: 'Will it create work scheduling problems when placed alongside existing projects?') Thus these questions introduce consideration of the nature of the firm-specific resources relevant to the project, another issue raised in earlier chapters.

Section D deals with financial data similar to that referred to earlier in this chapter.

Section E, dealing with the practicalities of putting the potential innovation into production, raises a new category of uncertainties. In the case of a radical innovation it will often be necessary to innovate production techniques in order to succeed. This magnifies the problems several times over. Even in cases where production techniques can be foreseen with reasonable security, it is still important to give consideration to overall capital and manpower requirements, as well as to do the obvious costing of production which is needed to feed in to the overall profitability estimates.

Finally in section F there are environmental criteria. The severity of the precautions required will be dependent on the industry, the nature of the legislation pertaining to the product and process technologies, and on the standards implied by the history or 'culture' of the company in question. This is clearly an area where calculation is difficult, and where responsibilities are open to argument. The treatment of the 'externalities' of technical change is still an underdeveloped area. It is nevertheless important to register that the role of political institutions can enter directly into the practice of evaluating alternative R & D projects. The role of such influences is considered in more detail in Part III of this book.

This discussion of the content of techniques for the evaluation of

R & D projects has shown that many of the issues which managers encounter are related to the broader issues of the structure and strategy of the R & D programme, as discussed above in Chapter 3. However, the uncertainty and the costs of collecting data on individual projects make the choice of projects to support an R & D strategy difficult and complex. It is therefore worth asking how valuable the techniques are in this task.

There is very little evidence on how widely formal evaluation techniques are used in R & D management. Many companies use an 'in-house' version which may owe a debt to some of the published techniques but is not highly formalised. Some companies probably do not use formal techniques at all. There is also some evidence that the evaluation procedures may be manipulated by the people involved in preparation of the data. Thomas (1970) for example has cited cases of engineers giving deliberately optimistic forecasts of R & D costs and times to senior managers for inclusion in selection procedures, in order to give pet projects a better chance of being funded. It might be thought that historic data on cost forecasts might provide a basis for a 'correction factor' to be incorporated in future forecasts – a sort of forecasting learning curve. But, as Freeman points out, the context of estimation will always be one of political advocacy and clash of interest groups, whatever the possibility for sober calculation (Freeman, 1982).

These comments suggest that there are serious limitations on the value of formal project evaluation techniques. However, as noted above, there may be a positive value in organising the collection of data, despite the uncertainty in the use to which it is put. This can usefully be expressed in terms of the Penrose argument concerning the role of firm-specific managerial resources. The performance of project evaluation may be consistent with the view that successful companies are efficient at converting managerial resources into a flow of managerial services. It also supports the view that the resources themselves can be augmented over time. These resources are used and developed by project evaluation procedures not only in the R & D department but in other parts of the firm. The habits of consultation, and the channels of communication between staff of different departments, which are fostered by the use of formal techniques, might atrophy or even fail to develop under a more *laissez-faire* regime. It has already been emphasised in Chapter 1, and supported by much of the case-study literature, that coordination between different parts of the innovating firm is an important contributor to the flow of innovative ideas, and to the success of innovations. Despite their informality and under-researched position in the literature, the existence of evaluation techniques in R & D management may therefore be seen as potential support to the behavioural accounts of the firm. Firstly, they illustrate the concept of managerial resources as both a source of, and constraint upon, the growth of a company. Secondly, the uncertainty and continual re-evaluation of projects, with its continual need to combine data from different sub-units in the company, gives support to Cyert and March's view that much decision-making is incremental in character. This is a reflection of the bounded rationality which characterises all business operations, but innovation and R & D in particular.

The assembly of a number of projects into a portfolio, after their submission to evaluation, has been seen to be affected by the strategic and structural aspects of the R & D programme, which derive in part the overall strategy of the firm. But the multiple sources of ideas for projects, and the uncertainty of the technical and commercial milieux from which they are drawn, constitute a permanent source of disequilibrating factors which continuously forces reconsideration of strategy and structure in the R & D programme and even at higher levels of the company. What results is an interaction between the flow of new project ideas and the set of received ideas, preferences, and commitments which make up the current strategy or direction of the firm. The balance between these two forces will shift back and forth and is a matter for investigation rather than prediction. The evolution of R & D portfolios over long periods, in different companies in the same industry, and across several industries, is therefore an important topic for further research in the study of technical change.

Monitoring and control

The initiation of R & D projects and assignment of budgets, personnel and targets, implies that these budgets and targets be monitored and controlled. The most obvious tool which might assist R & D managers in this task is the subsequent application of their evaluation techniques to existing projects as well as to new proposals. This would be consistent with the incremental nature of decision-making in this environment proposed in the previous discussion. As time passes, some of the information which was extremely uncertain at initial evaluation may become more certain, though it should be noted that it is in the nature of R & D to generate uncertainty as well as reduce it. After some time has passed it should be possible to determine whether any of the technical objectives of the project have been met. However, as Freeman (1982) notes, the important issue is often not whether something works or does not work, but how effectively it works, and at what cost.

The most favourable conditions for monitoring and control exist when a project plan can be drawn up at a fairly early stage. This allows milestone events to be nominated and their timing observed, it allows costs to be apportioned to particular tasks and components, and it allows scope for the use of operational research techniques such as network analysis (Twiss, 1980, Ch. 6). This gives the manager a more detailed picture of the progress of a project along a number of dimensions. It may enable use of parallel research strategies at some points to increase probabilities, followed by adjustment of staff allocation when one line succeeds. It also helps in the balancing of resources between projects, giving a finer adjustment mechanism for shifting personnel and equipment.

Even a simple project plan gives some scope for monitoring and control. The two key components in this activity are firstly the setting of targets in the form of specifications of performance, budgets and times, and secondly, the measurement of these parameters at various times. A common way to measure costs is to assign capitation rates to grades of staff and to require them to record time spent on different projects. Together with material and capital costs, this provides basic cost data. It then becomes possible to compare time-cost trends with plans. If some assessment of 'degree of completion' can be made, however notional, then progress-cost and progress-time comparisons can be drawn. Since there is some limited scope for time-cost trade-off, there can sometimes be scope for managerial action in the form of committing extra resources at key points to keep an element of a project on target for time, in order to avoid other related elements being delayed and incurring greater cost penalties. To some extent then, the familiar apparatus of managerial economics can play a part in monitoring and control.

It will be clear from this brief discussion however, that there is a continuing tension between the objective of control and the uncertainty and unpredictability of R & D work. If the unpredictability were unambiguously a bad thing for the organisation then this would strengthen the case for more effective control, but of course it is not so clear cut. The serendipitous nature of R & D work, and the possibility for a project to fail in its original purpose yet generate a result of considerable use in another context, is an important benefit of the work. Too rigid a control system would probably be dysfunctional in the long run. Hence we are forced to return to the nature of the specific managerial resources in the R & D department as the most useful variable in describing the control behaviour, rather than any formal analytical economic model.

4.3   THE ORGANISATION OF THE R & D FUNCTION

This is the last of the topics on the list of seven areas of decision-making which was introduced at the beginning of Chapter 3. We shall explore three issues in this context. First, the relationships between the R & D department and the other parts of the firm. Secondly, the internal organisation of the scientific personnel in the R & D department, and thirdly a brief consideration of the radical alternative of 'venturing'.

In Chapter 1 it was stressed that the inputs to the innovation process come from a variety of parts of the firm. This observation has been expanded in subsequent chapters. Thus the organisational problem is to achieve adequate flows of information between the various functional areas in the firm. The obstacles to this are numerous. One of the most substantial is the result of the very success of previous innovations, namely the proliferation of product groups, markets, and geographical areas in which a company may be active. This presents physical problems in arranging interactions between marketing and R & D personnel. More fundamentally, it may raise conflicts between the sub-units and within the sub-units themselves. (Compare the discussion of Cyert and March in Chapters 2 and 3). For example, if a complex product is being sold in a variety of different national markets, the incentives to develop new performance features on the product may not be the same in each of those markets. When the R & D projects which might satisfy these demands are discussed, it may emerge that the satisfaction of one requirement for improved performance may actually impose a penalty in another market. Thus the sales goals of the two marketing units conflict. If this is resolved by making the two improvements optional at the factory assembly level, the conflict shifts to the production unit, which will incur production regularity penalties, and possible cost penalties.

The resolution of problems of this sort within one product group is frequently attempted through committee structures which include personnel from all three of the major functional areas of R & D, market and production. As with any change in the division of labour, this may result in some new expertise in such coordination problems emerging amongst the personnel who participate in these discussions. Such augmentation of managerial resources is often explicitly recognised by the practice of seconding personnel from the functional areas to work in association with the other functional areas, or with semi-permanent coordination structures, for limited periods of time, and then returning them to the original functional area in the hope that their experience will assist the conduct of that function. At the limit, this practice may be fully formalised in a product planning department, separately organised from the R & D and marketing departments.

In the context of several product groups, the problem may be more acute. The question which arises in the case of an M-form company is particularly interesting. One alternative form of organisation is to decentralise R & D to the product divisions, which may maximise the contact between R & D and other functions in that product group. At the other extreme, some companies retain the R & D function at a centralised level, which may promote more interactions between different groups of scientists, and between R & D and strategic management, but may reduce contact with operating divisions. (See Green and Morphet, 1977 for further discussion and examples of these alternatives.) The choice of compromise level between these two alternatives has been analysed in economic terms by Kay (1982).

Kay's approach is to combine the transaction cost framework of Williamson (see Chapter 2 above) with the notion of synergy derived from the work of Ansoff. In essence he argues that there may be circumstances in which the decomposition of a U-form company into an M-form company may incur cost penalties if potential synergies within some functional areas are foregone. Thus there may be cases where the nature of the product groups and markets in which a firm is active will create great incentives to maintain all R & D in one unit, while diversifying production and marketing to market or geographical level. In other cases it may be more efficient to decentralise R & D and maintain a central marketing function. The contingent variables which Kay uses to control this degree of trade-off between synergies and the gains of the M-form organisation are the 'degree of relatedness' of the various activities of the firm, and the 'expected rate of technical change in the environment' (1982, Ch. 5). While this approach seems to be of some use in interpreting existing case studies, it is not clear how these two variables could be measured and used in any predictive sense.

While the analysis of organisational structures is obviously of importance both to the manager and to the student of industrial organisation, the role of individuals in achieving coordination in innovation should not be overlooked. Many of the most substantial contributions to the case study literature on innovation (Langrish *et al.*, 1972; and SPRU, 1972, for example) have found that particular individuals acting as 'product champions' or 'business innovators' have played key roles in the life of an R & D project or an innovation, even to the extent of circumventing the formal procedures and organisational structures designed by the senior management. This phenomenon is difficult to analyse in economic terms and, like a number of other topics in the study of innovation, it awaits the development of a sociology of the applied sciences before it can be better understood. Recruitment of scientific personnel with particular skills is another aspect of the importance of the individual which is noted in the work of Langrish, *et al.* This phenomenon has become more visible with the development of regional concentrations of new industries such as the microelectronics industry, with its volatile labour market. Analysis of such labour markets for R & D personnel could give useful insights into this aspect of innovation.

We turn now briefly to the second topic in this section: the organisation of R & D personnel within the R & D department.

The main issue which occupies the management literature on this topic is the relative merits and demerits of discipline-based, project-based and matrix organisation of the scientific and technical personnel (Twiss, 1980, Ch. 7). Discipline-based organisation has the merit of maintaining and developing special scientific competences, but has the demerit of complex work-flow problems, and the possibility of low commitment and lack of development of a commercial orientation. By contrast, project organisation may improve cohesion and commercial awareness, but restrict or even erode the specialist skills of those personnel who are deployed to a particular project for long periods. The solution of matrix management, in which these two structures are overlaid upon each other, is currently favoured in the literature (Loveday, 1984; Sbragia, 1984).

This fundamental problem arises because the functional characters of the corporate organisation and the reputational structures of science and technology (Whitley, 1985) are different. Scientific and technical personnel are likely to feel motivated to retain some degree of academic relationship to the field of research in which they work, and of course this is to the benefit of the company insofar as it acts as a source of extra ideas and competences (compare the discussion of basic research in Chapter 3). However, such activity is difficult to assess or control in the context of the short-term goals of a company. Thus the point of balance between cosmopolitan academic activity and internal corporate activity will be determined differently for different personnel and for different companies, by a wide range of factors. Once more, this topic depends upon an elaboration of the sociology of the applied sciences for more satisfactory discussion. While some work is available in this area, there is not space to discuss it here.

Faced with all the organisation difficulties associated with innovation which have been mentioned above, some companies may choose the more radical organisational solution of forming small specialist companies for the sole purpose of developing new products. This practice is sometimes known as venturing or New Business Development, and is the final topic to be examined in this section. The main reason why large companies may choose this option was perhaps identified before the practice even began, in the work of Burns and Stalker (1961). They noted that large firms tend to develop 'mechanistic' or bureaucratic management structures which are a constraint on the process of innovation. By contrast 'organic' management structures, which are often associated with smaller companies or business units, tend to be more informal in style, allow greater contact between functional areas, and be more conducive to innovation. It is this basic insight which the venturing strategy tries to develop.

Littler and Sweeting (1984), in an analysis of the experience of fourteen UK companies who have engaged in 'New Business Development' find however, that the culture and expectations of a parent company can sometimes continue to exercise a disturbing influence on the operation of the new business. A substantial proportion of their sample had discontinued the experiment after a few years. This experience may suggest that the gains of the 'organic' firm cannot be realised directly by the mature company, or it may be that the failings identified by Littler and Sweeting in the methods of the mature companies have corrupted the experiment from the outset. It may be that the more 'arms length' method of venture capital organisations' investing in autonomously started innovating firms may be more effective.

The root of this topic is, of course, the effect of firm size and market structure on propensity to innovate. This topic has a long history in the literature which has become more topical in recent years with the revived interest in small firms and entrepreneurship. Whilst the experiments with venturing and New Business Development are partly motivated by the desire to avoid some of the organisational problems of innovation in the established or mature company, it is clear that the current interest in small companies has deeper roots. The evidence on the relationships between firm size and innovation is a complex picture, which is deferred in this volume to Chapter 5.

4.4   CONCLUSION

The discussion of the organisation and execution of R & D in this chapter has revealed a variety of management practices which appear to depend on tacit knowledge, *ad hoc* procedures, and accumulated experience in particular fields. These practices are set in a context of technical and market uncertainty and organisational complexity, and depend upon the actions of scientists and technologists who are still subject to an attenuated form of the collegial control mechanisms of the scientific establishment. Not surprisingly, the formal algebraic apparatus of economic analysis is not very effective in this domain. Perhaps the most significant reason for this, apart from uncertainty, is that the basic unit of R & D is the *project*, which has a technical objective, and not the innovation, which has an economic dimension to its objectives. Thus the different level of aggregation of project and innovation forms the main obstacle to economic analysis. Nevertheless, several points of contact with behavioural theories of the firm have been made, though at a rather general level.

These characteristics of R & D and its management, which combine uniquely the features of firm, market and noncommercial research institution, reinforce the importance of seeing its product, technical change, as the outcome of the interactions of a number of different types of institution and practice. In discussing R & D in the firm, we have encountered the limits of the role of the firm in the pure sense. Parts II and III of this book consider the roles of market and political institution.

Part II
Economic Analysis and Technological Change

5 Patterns of Innovation

In previous chapters the problem of how innovative activities are managed within firms has been analysed. Reference has been made to different strands of literature dealing with theories of the firm, with its implications for the analysis of technological innovation, and with the way in which R & D is managed within firms. The theories of the firm which have previously been examined differ considerably with respect to their fundamental assumptions but they all share as their primary concern the behaviour of the firm. In other words the firm and not technological innovation is their unit of observation. In this chapter the focus of attention will be shifted away from the firm in two different directions: first, greater attention will be paid to innovation as a process, in which therefore common features may be shared in different firms, industries, countries and so on, and second, the analysis will often be conducted at higher levels of aggregation, such as the industrial sector. In this sense the chapter will still be concerned with microeconomic issues, since it will still be dealing with the composition of the economic system rather than with its aggregated flows, but at a level of aggregation higher than that of the firm.

It has already been observed that technology is considered exogenous to the economic system in the most simplified version of the neoclassical theory of the firm. Other theories of the firm, which differ in substantial ways from the neoclassical theory, do make a modest contribution to the problem of the origin of innovations. But there is still not a satisfactory explanation of the origin of innovations in these theories, and in addition there is a paradox in this lack of a satisfactory theory of the origin of innovations and the fact that R & D has been thoroughly institutionalised as a firm function. This contradiction is related to the tension between prescriptive and analytical literature (Chapter 1). The ideas and studies which will be examined in this chapter have been, implicitly or explicitly, an attempt to bridge this gap.

5.1   THE DEMAND PULL/TECHNOLOGY PUSH DEBATE I: THEORETICAL ANCESTRY

The work of two economists, Schumpeter and Schmookler is central to contemporary theories, yet paradoxically it could not easily be fitted into the theories of production and technological change which existed at the time they wrote. For both of them, although in quite different ways, technological change was a very important component of economic development. Their ideas have become very influential in shaping studies of innovation and technical change and have generated probably the four most important hypotheses about the origin of innovation.

Schumpeter was concerned with long-term economic development and structural change in capitalist societies. Entrepreneurs and monopolies occupy a central position in Schumpeter's ideas. The entrepreneur is the character in capitalist societies who discovers, often in an existing knowledge pool, ideas untried in commercial reality and introduces them into economic life. To achieve this the entrepreneur has to overcome barriers due to existing habits and institutions. This aspect of Schumpeter's ideas represents a radical departure from neoclassical economics. Not only are entrepreneurs and innovations incompatible with equilibrium, but the disequilibrium represented by the intrusion of innovations into the economic system, which Schumpeter calls 'the gale of creative destruction' is for him, the essential ingredient of capitalist economic development (Schumpeter, 1943). Radically new innovations lead to the emergence of completely new industries and create a renewed momentum for economic development. The supply of new technologies is, therefore, more important than the adaptation to existing patterns of demand. Furthermore, only product innovations can lead to the creation of new industries. They are thus more significant than process innovations, which can only lead to the increased efficiency of existing industries.

According to Schumpeter, an entrepreneur is motivated to run the risks inherent in introducing a new idea and in overcoming established barriers by the expectation of a temporary monopoly position, and the ability to enjoy large profits while the monopoly lasts, as a result of being the first to introduce a new idea. At least, this was the point of view of Schumpeter in his early works

(Schumpeter, 1934). In his later works Schumpeter (1943) shifted his emphasis from future monopoly expectations to existing monopolistic advantages as the essential factor in allowing the introduction of new ideas into economic life (Freeman *et al.*, 1982). He observed that innovation requires resources such as R & D and design which are expensive, have a minimum efficient size and sometimes show positive returns to scale. In a market of a given size a monopolist or an oligopolist can have easier access to these resources than an atomistic competitor. These two sets of ideas have sometimes been referred to as Schumpeter Mark I and Mark II. An important distinguishing feature between the two Schumpeterian viewpoints is that while inventive activity was entirely exogenous for Schumpeter Mark I it became at least partly endogenous for Schumpeter Mark II, since it was mostly conducted within large oligopolistic firms. Thus he acknowledged in his later work the growing institutionalisation of R & D.

In these ideas of Schumpeter there are the seeds of three of the four central questions or hypotheses about the origin of innovation.

First, whether the central character is the entrepreneur or the large firm, it is only by introducing radically new ideas into economic life that whole new industrial sectors can be generated. Technology, whether generated outside the economic system or in the large R & D laboratories of a monopolistic competitor, is for Schumpeter the leading engine of growth. Therefore the 'technology push' hypothesis of the origin of innovations finds a natural place in Schumpeter's ideas.

The second question is highlighted by the at least apparent contradiction between Schumpeter Mark I and Mark II. Will more and better innovations be produced by many entreprenuers or by few large oligopolists? Or, in other words, *what is the ideal market structure to stimulate innovation?*

The third question, for which Galbraith should perhaps be credited more than Schumpeter (Kamien, Schwartz, 1982), is related to firm size: *What is the best firm size to stimulate innovation?*

The fourth hypothesis is generally attributed to Schmookler (1966). He studied investment, stocks, employment and inventive activity in the railroad, petroleum refining, agriculture and papermaking industries in the USA from the first half of the nineteenth century to the 1950s. His time series, in the case of inventive activity, and of demand were of patents and investment in capital goods. Schmookler found that the time series for investment and patents showed a high degree of synchronicity, with the investment series tending to lead the patent series more often than the reverse. He found in particular that investment usually led the upswing from the troughs of economic fluctuations. On the basis of this evidence Schmookler argued that fluctuations in investment could be better explained by external events than by the course of invention and that, on the contrary, upswings in inventive activity responded to upswings in demand.

Schmookler did not argue that demand forces were the only determinants of inventive and innovative activity. If anything he was trying to correct the opposite imbalance according to which it was only the exogenous flow of inventions which could generate new investments and new economic activities. He used the example of the two blades of a pair of scissors to represent invention and demand as two interacting forces. However, and probably because he was trying to correct the opposite imbalance, the main emphasis on his work was on demand factors. In the literature he has almost always been referred to as the exponent of a demand-led theory of innovation.

The foregoing brief summary of Schumpeter's and Schmookler's ideas is designed to serve as an introduction to the theoretical background in which many important debates about the origin of innovation have developed. The demand pull/technology push debate and the market structure and firm size debates will now be described in more detail.

5.2   THE DEMAND PULL/TECHNOLOGY PUSH DEBATE II: EMPIRICAL STUDIES

During the Second World War science made considerable contributions to technological development. These contributions, exemplified chiefly but not only by the Manhattan project, had been in no small part responsible for the considerable increase in funding of R & D after World War II. In the immediate post-war period the belief that only pure science could generate a self-sustaining economic growth was expressed by many policy-makers (for example V. Bush, 1947). However, this belief started to be challenged during the late 1960s and early 1970s. A number of studies were carried out in order to ascertain the extent to which science had been responsible for technological development and economic growth. The slowing down of economic growth, especially in the USA, provided a further impetus for these studies. What was required was a detailed analysis of how the process of innovation took place in firms and industries. An analysis of this type could not be conducted within established disciplinary boundaries. Neo-classical economics dealt with technological change in an implicit way, making it almost impossible to understand how a shift between two different production functions took place. In the innovation studies of this period a more empirical approach was developed. One of the main features of these studies was the belief that any approach which studied a *large sample* of innovations might reveal some patterns or general laws which might aid understanding of this complex process. This could be called the natural history phase of innovation research.

It is impossible, given the space limits of this book, to describe in detail these studies. Table 5.1 gives a summary description of the best known of them as an introduction to the brief analysis attempted here. The reader interested in a more extensive account of them is referred to Mowery and Rosenberg (1979), Rothwell (1977) and Rothwell and Walsh (1979).

As can be seen from Table 5.1 the object of study and the unit of analysis of the various studies are very heterogeneous. They range from weapons systems in HINDSIGHT, to innovations in TRACES, to firms in *Carter and Williams* and *Wealth from Knowledge*, to industries in *Myers and Marquis* and SAPPHO. A concern with the mechanism of innovation is always present but the levels of aggregation of the studies are very different.

A great variety of methodological approaches was adopted in various studies. The principal ones were as follows:

The events approach

This approach is based on the assumption that an innovation is the result of a number of *events*, which can be classified as scientific,

TABLE 5.1 *Summary of innovation studies*

HINDSIGHT

An analysis of the relative contribution of basic science, applied science and technological events to the development of twenty weapons systems. Sponsored by the USA Department of Defence (DoD). (Sherwin, Isenson, 1967).

TRACES

An analysis of the relative contribution of non-mission research, mission-oriented research, development and application in the development of five innovations. Sponsored by the National Science Foundation (NSF). (TRACES, 1968).

Project SAPPHO

A comparative analysis of 'paired' successful and unsuccessful innovations in the chemical and scientific instrument industries. A total of 43 pairs was investigated (22 chemicals, 21 instruments) and the sample was international. The success/failure criteria were commercial. Also looked at factors associated with 34 of the failures. (Rothwell *et al.*, 1974).

The Hungarian SAPPHO

An adaptation of the SAPPHO pair comparison technique to 12 success/failure pairs in the Hungarian electronics industry. (Szakasits, 1974).

Carter and Williams

Studied the characteristics of 200 technically progressive firms in the UK. (A technically progressive firm was one which, on a necessarily subjective judgement, was keeping within a reasonable distance of the best current practice in the application of science and technology). (1957).

Myers and Marquis

Studied the characteristics of 567 successful technological innovations in five industries in the USA. (Railroad supplies, housing supplies, computer manufacturers, computer suppliers). (1969).

Queen's Award Study

Studied 84 innovations in the UK which gained the Queen's Award to

Industry for innovation between 1966 and 1969. Identified factors associated with success and factors causing delay in innovation. (Langrish *et al.*, 1972).

Belgian Study

Studied innovation strategy and product policy in 12 Belgian enterprises over a ten to fifteen year period. Success criterion commercial (profit margin better than 7 per cent). (Hayvaert, 1973).

Dutch Study

Studied the factors affecting the innovation potential of 45 Dutch companies in the metal-working sector between 1966 and 1971. The success criterion was commercial i.e.

$$\frac{1971 \ turnover \ of \ innovations \ marketed \ since \ 1966 \times 100}{total \ turnover \ 1971}$$

which gives a measure of relative innovative capacity within the sector. (Schock, 1974).

MIT Study

Investigated the factors affecting success and failure in innovation in five industries (automobiles, industrial chemicals, computers, consumer electronics, and textiles) and five countries (France, F. R. Germany, Netherlands, Japan, UK) – total sample consisted of 164 innovations. (Utterback *et al.*, 1975).

Textile Machinery Study

Investigated the factors surrounding the generation of 20 radical and 15 incremental innovations (all commercial successes) in the textile machinery industry. Looked also at the factors associated with 18 failures (10 incremental, 8 radical). The project included the detailed study of some 20 enterprises and the sample was international. (Rothwell, 1976 a and b).

Gibbons, Johnston Study

Compares the relative importance of different sources of information, one of which is the scientific community, as inputs to the innovation process. (Gibbons, Johnston, 1974).

Source:　Part of this table is taken from *Regulation and Innovation in the Chemical Industry*, R. Rothwell, V. Walsh (1979) end of table 5.4.

technological, mission-oriented and so on, and of which the relative contribution to the innovation can be assessed. The simplest way of assessing these relative contributions is to count the number of events in each category. This procedure, which has been followed in HINDSIGHT and TRACES is open to the objection that not all events are equivalent and so they should be weighed according to their intrinsic importance. Furthermore a decision has to be made about the length of time during which events are going to be considered. The choice of time period, ten years in HINDSIGHT, much longer in TRACES, can significantly affect the conclusions.

The process approach

In other studies innovation was implicitly or explicitly considered as a process, occurring through a number of stages (for example *Carter and Williams, Wealth from Knowledge*). But, there was no general agreement about the nature of the process. The dominant model of innovation available at the time was the *linear model of innovation*, according to which innovation is a sequence of stages, starting either from R & D or some perception of demand and ending with a product sold on the market. In its simplest form the linear model of innovation implied that each stage would be triggered by the output of the previous stage. More complex interactions or feedback processes were not systematically dealt with. However, this model was shown to be oversimplified by some of these studies, particularly in *Wealth from Knowledge* (Langrish *et al.*, 1972). In many cases information was gathered by means of interviews with personnel in the innovating firms. In the absence of an agreed model the factors judged to be crucial in determining success in innovation, and therefore the questions that were asked in the firms varied from study to study. Difficulties arise therefore in comparing the various studies.

Discriminating between success and failure

The comparison of successful and unsuccessful innovations in SAPPHO (Table 5.1) was based on information gathered on a number of factors judged to be potentially important in determining success in innovation. The presence or absence of these factors was tested in each of the pairs of firms studied. While some factors were associated with both successes and failures, other factors were only present in successful innovations.

Information inputs into the innovation process

Gibbons and Johnston assumed simply that innovation was a process which used information and tried to assess the relative importance of various sources of information.

These are just some examples of the great methodological variety of the studies summarised in Table 5.1. Straightforward comparability of their conclusions could, therefore, not be expected. Nevertheless these studies have generally been interpreted as suggesting that *demand* (or need) was the most important determinant of the innovation process (Mowery, Rosenberg, 1979).

Similarly, in their comparison of some of the previously quoted innovation studies Rothwell (1977) and Rothwell and Walsh (1979) find that in spite of the considerable methodological heterogeneity some similarities exist between the conclusions of the various studies. For example, two factors, *understanding of user needs* and *good communication and effective collaboration* tend to be strongly associated with success (or the lack of them associated with failure) in all the studies reviewed although the evaluation of such factors is not without its difficulties. A number of other factors such as efficient development work, quality of management, after-sales service and user education tend often to be correlated with success, though less highly than the first two. Rothwell concludes that although there is a considerable degree of agreement between the results of the various studies there are no easy explanations or panaceas which can be offered to management. The successful innovator 'must take care in all the areas of competence encompassed by the innovation process'. (Rothwell, 1977).

Therefore, some agreement exists amongst the conclusions of various studies. But is it enough to justify the general conclusion that demand (or needs) are the most important factor contributing to the success of an innovation? According to Mowery and Rosenberg (1979) this conclusion is not warranted in view of some methodological short-comings and the lack of comparability of the various studies. They argue that the loosely-defined concept of 'needs', and not the narrower and more rigorously-defined concept of demand, was frequently the independent variable, making a 'need-pull' conclusion more likely. Furthermore, technology-push factors were often, in their view, rendered impossible to find in the case studies by being conceived in a caricatured way as entirely *scientific* events free of any economic component whatsoever. In a world of institutionalised R & D such events are rare. Both of these conceptual filters tend to bias interpretation in favour of 'need-pull' conclusions.

On the basis of their critical analysis of the various studies, Mowery and Rosenberg therefore conclude that both supply and demand are important determinants of success in innovation. They pointed out that not only are supply and demand both important, but that the coupling of technology and market is essential if an innovation is to be successful. This might not seem a very original conclusion to come from a large number of studies carried out by highly qualified research teams. However, these studies have made a fundamental contribution to our understanding of the innovation process in at least two ways: first, they have shown that innovation is a very complex process and that, therefore, it is not possible any more to consider any particular factor be it science or user needs as the sole or the fundamental determinant of innovation; second, the wealth of empirical evidence gathered in these studies has become both a challenge to established economic theories and an opportunity to build alternative ones. The demand-pull/technology-push debate has not been settled but has been advanced by these studies to a qualitatively different level. In particular, the introduction of more work based on particular industrial sectors, on international comparisons, on trade flows, and the more explicit use of the theoretical ideas deriving from Schumpeter and Schmookler, has taken advantage of the elaboration and detail of the demand-pull/technology-push debate. Important examples of this approach have come from the work of the Science Policy Research Unit.

For example, Pavitt (1980) analyses the innovative performance of Britain and of its main international competitors in a number of industries such as textile machinery, shipbuilding, coalmining machinery, forklift trucks, steel, electrical plants, nuclear reactors and semiconductors. With notable exceptions, for example in coalmining machinery, the UK is found to have lost world market shares to established competitors such as West Germany, and to relative newcomers such as Japan. These detailed studies of innovation in particular industrial sectors are combined with studies on long-term trends in Britain's industrial performance, and on the relationship between innovative activities and export shares. From the viewpoint of the demand-pull/technology-push debate this work is interesting in two respects. First, relative rates of patenting in the United States by various countries are used as a proxy for the innovative activities underlying their exports and they are correlated with export shares in the US market. In a large number of industries a very high correlation is found between patenting and export shares in the USA. The conclusions found using patents confirm those from studies of particular industrial sectors: Britain is lagging behind her most important competitors. Second, the relative decline of Britain as a technological power is interpreted as being not only the consequence but partly as a cause of a more general economic decline. The policy implications of this change of perspective are considerable: an active innovation policy could be an instrument to reverse this decline. Clearly, although without espousing mono-causal theories of innovation, technology-push is given a far greater weight here than in the previous studies.

In more recent work at SPRU, Freeman *et al.* (1982) have been mainly concerned with the relationship between technical change and long-term structural economic changes. This topic will be dealt with in considerable detail in Chapter 7. However, the discussion of the development of the synthetic materials, chemical and electronics industries, based on previous SPRU studies (Walsh *et al.*, 1979, Walsh, 1984, Dosi, 1981) makes an important contribution to the demand-pull/technology-push debate, and it is worth summarising it here. Patterns of investment, scientific papers, patents, innovations and output are analysed for both industries. Neither technology-push nor demand-pull are found to predominate systematically but each one of them can lead the other at different stages in the development of the industry. If any generalisation can be made, technology-push tends to be relatively more important in the early stages of development of the industry while demand-pull tends to increase in relative importance in the mature stages of the product cycle (Freeman *et al* 1982). Therefore neither the ideas of Schmookler nor of Schumpeter are adequate alone. A combination of their ideas seems to explain better the development of these industries.

5.3   INDUCED INNOVATION

We have seen that a common conclusion of the studies of innovation reported in the previous section is that the influence of demand side factors on successful innovations can be very great. Indeed conscious attention to user needs on the part of the innovator is often emphasised as a necessary feature of success. It is a logical progression from this observation to consider the possibility that a general theory of innovation should be able to relate directly the rate and direction of innovative activity to the structure of demand. We have seen in Chapter 2 that neoclassical theory of the firm can only provide a limited scope for this connection: in which changes in factor prices affect choice within a range of *existing* techniques. Managerial theories of the firm, however, were found to show some promise in this respect through development of the concept of technological opportunities and varying returns to R & D in different technical fields.

These approaches however start from the level of the firm and the individual innovation, and require generalisation to more aggregated levels in order to make statements about patterns of innovation. This chapter is concerned with work which has *started* from the observation of patterns of innovation and made conclusions for other levels of aggregation. There has been a considerable tradition of theories of *induced innovation* which have developed in this way relatively independently of developments in theory of the firm. The original formulation of this theory is due to Hicks (1932):

> A change in the relative prices of the factors of production is itself a spur to invention, and to invention of a particular kind – directed to economising the use of a factor which has become relatively expensive.

However, since this original statement, the theory of induced innovation has been developed for a variety of purposes, not all of which are relevant to this book. A brief description of these developments will be given here. The reader interested in a more detailed account is referred to Binswanger and Ruttan (1978).

As we have already seen (p. 26) changes in technology can be neutral or have a factor saving bias. For example $Q_1 \rightarrow Q_3$ in fig.

2.2 would be an example of labour-saving technological change. If, for descriptive purposes only, we assume that Q_1 and Q_3 are production functions for a national economy at different times then more capital and less labour would be employed to produce a constant output in the new situation. At first sight it might be thought that this change would cause unemployment but this is not necessarily the case. For example, total output could increase enough to maintain constant employment. However, it is understandable that the previous question has worried economists, particularly during periods of depression when output can even decline. The suspicion arose that technological change might have an *intrinsic labour-saving bias* which would in the long term cause widespread unemployment in industrial societies. Doubts about the validity of the induced innovation hypothesis have been raised by Salter (1966) who answered the previous question by saying that technological change has no intrinsic factor saving bias, but that entrepreneurs tend to reduce all their costs, not just labour costs, and that innovations are generated and adopted accordingly.

Kennedy (1964), von Weizsacker (1966) and Drandakis and Phelps (1966) developed another approach which enables one to derive theoretically the induced innovation hypothesis. Given the existence of a production function they assume that the entrepreneur maximises the current rate of unit cost reduction subject to a production–possibility frontier, which determines the possible combinations of improvements in the efficiency of the factors of production (for example labour and capital). Through formal analysis they find that there is a greater tendency to save a particular factor, the greater is its share. The answer therefore differs from that of Hicks, since now the inducement is proportional to a *factor's share* and not to its price, and from that of Salter since the contribution to cost reduction of different factors can be very different depending on their share.

Two things can be noted from this brief description of some approaches to induced innovation: first, inventive and innovative activity have become at least partly endogenous to the economic system. Secondly, however, these approaches have been conceived at a highly aggregated level and cannot therefore lead directly to a clear understanding of innovation mechanisms and to the formulation of innovation policies.

Some further difficulties with the highly-aggregated approach of Kennedy, von Weizsacker, Drandakis and Phelps can be illustrated briefly. For example in this approach, inventive and innovative activities do not involve the allocation of any resources and therefore any costs. This in turn implies that new techniques are instantly available and therefore that no diffusion problem exists (Stoneman, 1983). A paradox has arisen in attempting to apply the approach to the findings of Habbakuk on the development of US and British technology during the nineteenth century. According to Habbakuk, US technology during the nineteenth century had become more capital-intensive than UK technology in an attempt to save labour, which was more scarce and therefore more expensive in the US (Habbakuk, 1962). Labour was supposed to be more expensive in the US due to its greater abundance of land, which would have determined a higher marginal product of labour in agriculture and consequently higher wages. However, according to Stoneman (1983) it is possible to explain Habbakuk's findings by means of this approach only under restrictive assumptions in which factors of production such as labour and capital are specific to agriculture and industry respectively. In other words, Habbakuk's findings cannot be given a *general* explanation within this approach and the restrictive conditions under which they can be explained do not always seem to be realistic. For example land and capital are used as inputs both in agriculture and in industry.

An alternative explanation of the same findings has been provided by David (1975), who begins from the assumption that a lot of technological learning takes place in the form of *learning by doing* (Arrow, 1962). From this basis it is argued that an initial choice of a given technique leads to learning around the technique itself and consequently to the development of similar techniques. According to David, capital-intensive techniques tend to require larger amounts of raw materials per unit of output. In the US, which has a greater abundance of raw materials, an initial choice of more capital-intensive techniques could have been subsequently reinforced by 'learning by doing'. David's explanation constitutes an alternative induced innovation approach which, by relying on learning by doing, and therefore on experience, creates a more intimate link between the origin and the diffusion of a given technology. Furthermore the factor of production which is proposed as inducing changes in technology is in this case raw materials and not labour or capital.

At this point it is interesting to recall the discussion of inducement provided by Rosenberg, which was outlined in Chapter 2. In his approach there is a much more complex environment in which a variety of inducements, such as scarcity of raw materials, bottlenecks in technological developments and political factors can play a part. Changes leading to a greater efficiency in a part of an integrated technological system can create an efficiency bottleneck in the parts of the system in which there has been no change. This imbalance will induce innovation in the parts of the system which have remained relatively less efficient. The classical example of this inducement mechanism is the development of the textile industry, in which the increasing efficiency in spinning, due to innovations in that sector, led to a productivity bottleneck in weaving. This was in turn remedied by innovations increasing the efficiency of weaving. Rosenberg also supplies evidence of a labour-saving bias existing in the British textile industry in the nineteenth century. Industrialists sponsored labour-saving innovations in order to avoid strikes and to overcome labour's reluctance to accept their conditions. A third type of inducement is due to the increasing scarcity of raw materials which are important inputs to industrial processes. Examples such as France's early commercial leadership in the synthetic alkali industry, or the development of the nitrogen fixation process in Germany are seen as closely related to wartime shortages of Spanish barilla and Chilean nitrates respectively. (Rosenberg, 1976).

These contributions of Rosenberg and David illustrate a general point concerning the success of theories of induced innovation. This is that they are more convincing and susceptible to testing and development to the extent that they incorporate an explicit microeconomic component. The earlier theories did not do this since they were more concerned with potential macroeconomic effects of factor saving bias in induced innovation. The microeconomic thrust which has been developed more latterly in this field provides the possibility of a more direct link to the theory of the firm and the treatment of technical change within that theory: as discussed in Chapters 2 and 3 of this book. The work of Binswanger and Ruttan (1978) is perhaps the clearest example of a theory of induced innovation which has substantial overlap with the other microeconomic work on innovation.

Their approach aims at constructing a theory of induced innovation which can describe how research and development is actually carried out. Firms have a choice of research activities each of which affects the factor intensity characteristics of the production process in a different way. Relative changes in factor prices will determine which projects are most convenient within the research portfolio of a given firm. The convenience of a project is determined by its contribution to reducing costs. A project which is optimal under a given set of factor prices will become less convenient than another one if, for example, wage rates increase. However, the most convenient or appropriate project might not be the easiest to achieve or, in other words, savings in factor costs could be outweighed by the higher cost of the project itself. Progress is, in any case, always limited by a scientific and technological frontier, the most advanced knowledge existing at a given time. This frontier is usually not even reached, since diminishing returns to R & D can in general be assumed. A research manager is, therefore, faced with a choice determined by three main factors: the relative contribution to reducing production costs of different research projects, the cost or intrinsic difficulty of the project itself and the extent to which a particular research project can be pushed towards its scientific and technological frontier. This means that changes in factor costs will not always have entirely predictable effects on the bias of technological change, which is a different outcome from macroeconomic induced innovation theory.

It will be clear to the reader that this approach of Binswanger and Ruttan, which presents inducement in terms of benefit of the innovation, cost of the innovation, and limits of the technical frontier, is very close indeed to the treatment of invention costs by Stoneman reported on p. 45 of Chapter 2, and to the treatment of determination of R & D budget levels reported on p. 55 of Chapter 3. In those cases formal models were used to express the results of behaviour which had been deduced from managerial theories of the firm. In the case of Binswanger and Ruttan, we have arrived at the result through examining the evolution of theories of induced innovation toward a more explicit microeconomic content. This would appear to represent a genuine and fruitful convergence in the economic literature on innovation.

5.4 INNOVATION, FIRM SIZE AND MARKET STRUCTURE

The 'demand-pull' and 'technology-push' hypotheses are closely related to the 'intrinsic' features of the innovation process. The second and third Schumpeterian questions, on the other hand, are related to the nature of the institutions within which innovations are created. Is a monopolistic or a perfectly competitive market the more conducive to innovation? Are large or small firms the more innovative? These two questions are similar but not coincident: although larger firms are frequently found in more concentrated markets it is sometimes possible, by comparing for example different industries or different national markets, to find larger firms in less concentrated markets.

Monopoly was considered by classical and neoclassical economists to be an inefficient form of market organisation, leading to restrictions in output and to higher prices than in a perfectly competitive market. But Schumpeter pointed out that this was true only if competition occurred exclusively in terms of price (Schumpeter, 1943). Many firms, on the other hand, compete by developing new products and this type of competition is substantially different from that based on price. He argued that the resources required to develop new products could only come from the supernormal profits possible in a monopolistic market. Therefore a perfectly competitive market could lead to the highest efficiency only in a static world, in which a constant set of products and services is produced at costs continuously decreasing over time. According to Schumpeter, firms maximising this *static* efficiency make it impossible to develop new products. Monopolies and oligopolies, on the other hand, while less efficient in a static sense, allow the introduction of new products leading to a greater *dynamic* efficiency.

According to this view therefore, a monopolist would have a series of advantages over a perfect atomistic competitor which would make it easier to develop or adopt innovations. For example, a monopolist could prevent imitation of an innovation by competitors by a number of means ranging from patents, copyright and trade marks to the control of distribution channels. Also, since internal funding is an aid to secrecy in innovation, a monopolist would have a considerable advantage in this respect as well. Furthermore, a monopolist with an established reputation in R & D might more easily attract creative people. (Kamien, Schwartz, 1982).

Monopolies, however, do have some very well known disadvantages. The supernormal profits and greater resources that monopoly power can provide may either be used to innovate or to survive without innovating. One could argue that not innovating makes long-term survival more difficult, but there are examples of monopolists using their monopoly power in less than innovative ways. For example, firms in very concentrated markets often become very bureaucratised and unable to develop certain types of innovation (Kamien, Schwartz, 1982).

Economists have traditionally assumed that monopolies can survive without innovation as a result of lack of competition. According to Schumpeter (1943) on the other hand, whilst price competition might be absent, other types of competition largely based on the appearance of new products might affect a monopoly. For example a new technology (such as digital watches) might threaten established market positions based on a previously dominant technology (for example mechanical watches). However, since there are examples of both innovative and non-innovative monopolies and oligopolies, the possibility of non-competitive monopolies cannot be ruled out.

The degree of monopoly existing in a given market will not only influence the rate of creation of innovations but also the rate at which these innovations are adopted by other innovators and users. For example, by extending in time the monopoly of an innovator, imitation costs are caused to increase and the rate of diffusion of the innovation is consequently slowed down. It is this delicate balance between incentives to produce and incentives to utilise technological innovations, that all the systems for the protection of intellectual property rights (such as patents, copyrights, registered designs and so on) try to achieve.

The reasons for expecting firm size to influence innovative activity are similar to those previously examined for market struc-

ture. For example if, as Galbraith thinks (1967) the cost of innovation is continuously increasing over time, large firms would be in
an increasingly advantageous position. Other possible advantages
of large size are economies of scale in R & D resulting from such
factors as an environment in which researchers can share their
findings with more colleagues and greater chances to exploit
serendipitous discoveries (Kamien, Schwartz, 1982). There is,
however, evidence that large firms can have disadvantages as well
as advantages. For example unexpected research findings, rather
than being better exploited, could be wasted due to poor communications or to lack of management interest; or researchers'
motivation might be lower than in a small firm because compensation may be less directly related to their performance. There are
many examples of entrepreneurs leaving established large firms to
set up their own more highly innovative small firms. In this case as
in the market structure case, both theory and casual observation
lead to arguments both in favour and against the hypotheses.

In addition to theoretical analysis, there have been many attempts to test these two hypotheses empirically. The sheer volume
of this work prevents a comprehensive discussion in this book and
for this purpose the reader is referred to more specialised literature (Kamien and Schwartz, 1982; Scherer, 1980). However, the
problems encountered in these tests and some of the important
conclusions will be summarised here.

For the purpose of testing them empirically the two Schumpeterian hypotheses can be reformulated as follows:

1. Research intensity increases more than proportionally to market concentration.
2. Research intensity increases more than proportionally to firm
   size (where research intensity means the ratio of some measure
   of a firm's research activities to total firm activities, for example R & D expenditures divided by firm turnover).

In the vast majority of cases, the empirical test took the form of a
statistical correlation between some measure of research intensity
and some measure of either market concentration or firm size.
There are a number of difficulties which complicate both the
execution and the interpretation of these empirical tests. Firstly,
not only is information generally not available on a systematic
basis, but the information which is available does not always

capture the relevant dimensions of the phenomena under investigation. For example, the research effort of a firm can be represented by means of R & D inputs, such as R & D personnel or R & D expenditures. However, contributions to innovation coming from departments other than R & D, such as production or marketing, are not captured by these measures, although sometimes these contributions are likely to be very important.

Apart from these difficulties the use of R & D inputs to measure the research intensity of a firm is only justified if research results are related to research effort. According to Kamien and Schwartz (1982) currently available findings seem to imply that research results are approximately proportional to research effort, but research in this area is still under-developed (OECD, 1982) and no definitive conclusions can be made. It is extremely difficult to define all the inputs and outputs of innovation. The appearance of new products is not always easy to distinguish from marginal modifications of existing products. Patents have been used as a proxy for the output of innovative activity but, in spite of the easy availability of information about them some problems, such as the comparability of patents of different 'quality' or the wide variations in the roles that patents play in different industries, restrict their use as indicators of the output of innovation.

On the other side of the supposed correlations, substantial difficulties are encountered in measuring market concentration and firm size. In both cases more than one type of measurement is available but their results are not always in agreement. For example, market concentration is usually measured either by means of indicators which compare actual prices of goods with the prices which would prevail under perfect competition or by indicators based on market structure, such as the four firms concentration ratio (the percentage of industry output supplied by the largest four firms). Strictly speaking the first type of indicator measures monopoly power rather than market concentration, and they are not necessarily proportional. (Kamien, Schwartz, 1982). Firm size can be measured by employment, turnover, assets and so on, but different measures can sometimes lead to substantially different results in the analysis. Part of the uncertainty in the interpretation of research in this area is due to these difficulties of measurement and definition.

Other difficulties of interpretation are conceptual. For example, if scale economies exist in R & D, large firms could derive greater benefits than small firms by allocating the same percentage of their turnover to R & D. Clear evidence for these scale economies exists only for the chemical industry (Kamien, Schwartz, 1982) but the problem cannot be dismissed *a priori*. However, perhaps the most serious interpretation problems are due to the fact that variations in R & D intensity between different industries can be larger than inter-firm differences in the same industry (Scherer, 1980). These findings cast some doubt about whether firm size and market structure or technological opportunity are the primary variables of the problem.

Given the presence of these uncertainties it is hardly surprising that in spite of the existence of a vast literature on the relationship between market structure and innovation no firm conclusions can be drawn. The only exception is the chemical industry for which there is considerable international evidence of a threshold in both market structure and firm size below which little innovative activity takes place (Kamien, Schwartz, 1982). For other industries findings are much more contradictory and the Schumpeter Mark II hypothesis has so far not been confirmed. Slightly clearer conclusions can be drawn about the relationship between firm size and innovative activity. The consensus is that there is a tendency for R & D intensity to increase with firm size up to a certain size, and then to decrease. (Karmien, Schwartz). This implies that there is an intermediate firm size which represents the best compromise between the advantages and the disadvantages of large firms. However, this consensus has been challenged by Soete (1979) who does not accept the conclusion that R & D intensity might decrease beyond a certain firm size and argues against rejection of the Schumpeter Mark II hypothesis. Thus, firm conclusions about the relationship between firm size and innovation are also ruled out by the existing literature of empirical studies.

Indeed, we are convinced that the existing confusion is the result of inadequacies in the theoretical framework used in these analyses (and that the solution does not lie in more empirical work). In particular, we question the implicit assumption that economic variables such as firm size and market concentration are the *independent* variables of the problem, that is, the *causes* of different patterns of innovation.

Many studies show (as we have previously noted) that

inter-industry differences in technological variables such as R & D spending or rates of patenting are as significant, if not more significant, than inter-firm differences in the same industry. The explanation proposed is that industries vary in their degree of *technological opportunity* and that industries with greater technological opportunity will tend to show a higher research intensity and a greater degree of technical progress.

The existence of differences in technological opportunity between industries has the following implications for the theoretical framework mentioned above: technological opportunity, in addition to differences in firm size or market structure could become another, (or even the major) cause of different patterns of innovation; or technological opportunity as the *primary* variable in the relationship and the initial advantage of the first-comers in a new sector of high technological opportunity may subsequently lead to the establishment of an oligopoly.

The latter conclusion amounts to a reversal of the direction of causality – from market structure to innovation – which had been implicitly or explicitly assumed in all the previous studies (Momigliano, Dosi, 1983). Indeed, this was the conclusion we anticipated in our discussion of technological opportunity in Chapter 2 (p. 00). Such an assumption leads to a merging of studies of industrial organisation and of patterns of innovation and will be discussed at greater length in the following sections.

5.5 TECHNOLOGY, THE SELECTION ENVIRONMENT AND THE FIRM

A theory which could in principle *model* the processes of technological innovation and technological change should be able to provide a realistic and accurate representation of technology, of the firm, which in a more general sense would be the institution where innovations are generated and adopted, and of the selection environment in which the firm operates. These three – technology, the firm and the selection environment – can be considered analytically separable but interacting.

Several theories of the firm have previously been examined. They range from neoclassical to behavioural to managerial theories of the firm. Two aspects in which they differ and which are relevant for the purposes of this section are information and uncertainty. Innovation is a phenomenon which escapes the neo-classical assumption of perfect information because it creates more uncertainty and more information. The more radical an innovation the more information about it will be available only after the innovation has been developed. Moreover, learning effects, the existence of which have been unequivocally demonstrated, imply that the availability of information about a given technology becomes progressively less imperfect during the period of adoption and use of the technology itself.

There is therefore a basic contradiction between innovation as an information-generating activity and the assumption that information is available at the outset. Furthermore, even when information exists, access to it is certainly not unrestricted and varies widely among firms. Indeed, the differential access to information and knowledge is likely to be one of the most important variables determining firm success. From the viewpoint of the analysis of technological innovation, firm behaviour can be described as decision-making under conditions of imperfect information and uncertainty. We have seen in Part I of this book that behavioural theories of the firm have much to offer to the analysis of technological innovation. We saw that they allowed the development of models of innovation as a component of the discretionary behaviour of managers, which results from intrinsic characteristics of firms, from characteristics of technological opportunities, and from structural features of the industry and the market. Moreover there is still the possibility of a 'residual' degree of indeterminancy in innovative behaviour which is captured in Freeman's notion of innovation strategies, and which clearly also results from the intrinsic uncertainty of innovation. A particularly powerful attempt to synthesise some of these factors into one model is contained in the work of Nelson and Winter (1974). Their model combines many of the elements discussed above, while going on to deduce the consequences resulting from the interactions between firms behaving according to the model, and thus generating a novel theory of economic growth resulting from technical change. This latter dimension of their work is dealt with in Chapter 6, but this is an appropriate point to consider the microeconomic part of their model.

They start in a manner reminiscent of Cyert and March (1963) by postulating that firms have *decision rules* which can only be observed and not derived from profit maximisation or other neo-classical assumptions. However, according to Nelson and Winter, firms' decision rules do not change continuously but are character-ised by *short run stability*. In other words firms tend, for example, to use the same production techniques for considerable periods of time. Change does take place however, and therefore the con-stancy of decision rules can only be a good approximation on a short-term basis. Whilst adhering to constant decision rules firms continuously carry out 'goal oriented search processes', such as R & D or operations research. These search processes have to be modelled as a fundamental component of firm behaviour by speci-fying, for example, what goals of the firm are operative in the search process and what is the field of search. Indeed, Nelson and Winter define innovation as a change of decision rules and say that such a change in decision rules is more likely to be stimulated by threats and adversities than by a situation characterised by favour-able outcomes. Nelson and Winter's 'evolutionary' model of firm behaviour is represented in Figure 5.1.

We have already seen in the discussion of Kay's work in Chap-ters 2 and 3 that there is ample reason to accept the existence of these stable decision rules and managerial preferences with respect to the more aggregated aspects of innovative behaviour, such as R & D budgeting. In the case of the more disaggregated aspects such as the directions and objectives of particular projects and groups of projects, we argued in Chapters 2 and 3 that the concept of technological opportunity, and the related concept of techno-logical trajectory were powerful concepts for the explanation of decisions. This is the next element in the Nelson and Winter model.

Technological trajectories can be thought of as the disaggre-gated decision rules that firms adopt, and change from time to time, concerning the detail characteristics of their products and processes. Technological decision rules will not only be character-ised by short run stability but there is evidence that they may be common to a very large number of firms or to whole industrial sectors. In other words each firm will not only adopt technological solutions and approaches which are going to remain stable for short periods of time, but which are also similar to those adopted by other firms operating in the same technology. At least, this is

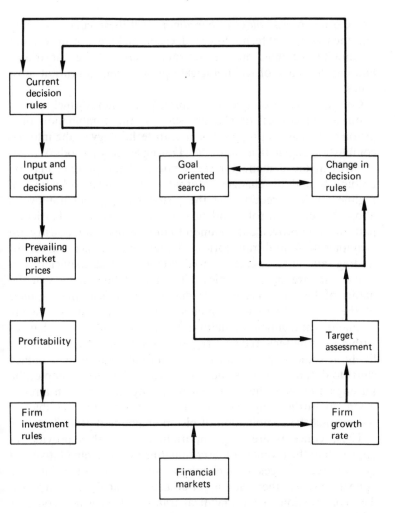

FIGURE 5.1   *Nelson and Winter's evolutionary model of firm behaviour*

likely to happen after a certain stage during the process of maturation of a technology. Utterback and Abernathy (1975) have introduced the concept of *dominant design*, which emerges as an
industry proceeds from the initial stages, characterised by a multiplicity of designs, towards maturity; Sahal (1981) has spoken of
technological guideposts; and Dosi (1982) has compared natural

trajectories to *technological paradigms*. All these concepts imply that the freedom of technological decision-making of firms will not be infinite but that, on the contrary, there will be constraints limiting the range of technological options open to firms at most times.

Consider some examples of technological systems which exhibit 'natural trajectories' of change, such as the combination of a piston engine and a propeller in aircraft technology or the internal combustion engine in a motor car. During lengthy periods of time, changes in the demand for the services of the product may be accommodated by changes internal to the given technological regime (such as redesigning the piston engine) and not by an altogether different solution (such as a turbine engine). Particular patterns of improvement common to most firms in an industry are examples of natural trajectories. The tendency to increase the lift to drag ratio in an aircraft, or to decrease the drag coefficient in a motor car, are such examples. Therefore, while changes in the nature of demand do exert inducements on the firm, the response of the firm to these inducements will be mediated by existing technological regimes and natural trajectories. Of course, changes within existing technological regimes and natural trajectories are likely at some stage to run into diminishing returns and induce shifts to different regimes and trajectories. Thus, for example, the jet engine became the dominant technological regime in aircraft when no further improvements in speed appeared possible by redesigning the combination of piston engine and propeller.

These concepts are very important because they provide an approach to the problem of discriminating between qualitative and quantitative changes in technology. In spite of their intuitive appeal however, they cannot yet be used for analytical purposes. The construction of an analytical framework in which concepts such as natural trajectories and technological paradigms can be accommodated can benefit from research aimed at *mapping* and *measuring* technological change. An approach common to a number of authors both at a theoretical and at an empirical level has been to consider technology as multidimensional and therefore represented by a number of characteristics. (Lancaster, 1966; Griliches, 1971; Martino, 1983). Furthermore it is possible to distinguish between characteristics which describe the internal structure of the technology and characteristics which describe the services performed to the users of the technology. Saviotti and Metcalfe (1984) have called the former *technical* characteristics and the latter *service* characteristics, arguing that there must be a pattern of mapping between the two types of characteristics. The emergence of a new technological regime can then be represented by entirely new characteristics, while more incremental types of changes lead only to different values of the same characteristics. Furthermore a plot of the N characteristics representing a technology at different times provides a *map* of the evolution of the technology. In this way Saviotti and Bowman (1984) have shown that at least two distinct trajectories becoming gradually more separated are evident in the evolution of aircraft technology. Research based on a multicharacteristics description of a technology is at a very early phase, but it already shows potential for the testing and applying of concepts such as technological regimes and natural trajectories.

The examples of natural trajectories and of technological regimes which have so far been described are specific to particular technologies. Nelson and Winter also give examples of natural trajectories which are common to large numbers of industries over long periods of time. Mechanisation, the exploitation of latent scale economies and the tendency to an increasing use of electricity are examples of such more general trajectories. These can be seen as giving further support for the basic proposition being put forward. This is given that the variety of technological responses of firms is in general not as great as the variety of stimuli coming from the economic environment or, in more familiar terminology, a change in demand will only rarely determine a shift of technological regime, paradigm or dominant design.

In addition to the firm and to the technology, the third element required for the description of the process of technological change is the environment in which the firm operates. In a capitalist economy the environment will most often be a market. From the market the firm purchases raw materials, labour, capital goods, technology and so on, and to the market it sells its output in competition with other firms. We have seen however, that the growing complexity of market and hierarchical forms of coordination means that many innovation decisions may not be taken in the context of a pure market. Nelson and Winter (1977) have proposed the concept of *selection environment* as a more general analogue of a market in which producing institutions compete.

In each type of selection environment there will be different motivations, rewards and criteria for success. For example, the behaviour of an institution in the public sector providing a health service (such as a hospital) will not be determined by profit maximization or growth, in the same way as for a private firm. Thus rules for competition vary in different selection environments.

In this way the model of Nelson and Winter goes some way to explaining some of the patterns observed in innovations, which have been the subject of this chapter. Their work represents a general tendency in the literature to combine microeconomic inducement mechanisms, managerial models of firm behaviour, the structured fields of technological possibilities which are common to large numbers of firms but which are not sufficiently powerful to generate identical responses from firms due to the complexity of selection environments, and uncertainty. This latter aspect of the argument has the added consequence that we can expect discontinuities when natural trajectories are abandoned and new ones attempted. This is of some importance to the issues of industrial and market structure, as was discussed in Section 5.4, and *a fortiori* to the question of structural change which is discussed in Chapter 7.

However, so far the analyses of technological innovation which have been described relate mainly to the stage of the *origin* of innovation. But, if the creation of patterns and constraints within the evolution of a particular technology occurs when the technology itself is *maturing*, then it is clear that the origin and the diffusion of a technology are not entirely analytically separable. The process of maturation necessarily implies the diffusion of the boundaries which constrain the subsequent evolution. Therefore it is now necessary to examine the theories of the diffusion of technological innovation.

5.6 THE DIFFUSION OF TECHNOLOGICAL INNOVATION

The diffusion of innovations not only spreads throughout the economy the increasing productivity and other benefits resulting from the innovations, but it also conveys information about their performance both to potential adopters and to manufacturers of the innovations themselves. Although there is clear evidence that these two stages of the origin and of the diffusion are interacting it is possible to separate them analytically. Indeed this is what has been done either implicitly or explicitly in most of the studies analysed so far. Most of them were concerned with innovation rather than diffusion. In this section a number of approaches will be reviewed which are concerned almost exclusively with the stage of diffusion.

The fundamental elements in the process of diffusion are the innovation which diffuses, the population of potential adopters and their process of decision-making, and the flows of information about the innovation between manufacturers and adopters. The models of diffusion which will now be reviewed differ with respect to the way they represent these elements.

The epidemics model

In an epidemic a disease is transmitted to healthy individuals by contact with infected ones. As the disease spreads the number of carriers increases and so does the rate of diffusion until the number of healthy individuals left becomes so small that the rate of propagation has to decrease. In the case of the diffusion of technological innovation what diffuses is information about the innovation itself. Firms will have very little information, if any at all, about an innovation that has not yet been widely adopted and will, therefore, associate with it a high degree of risk. As more firms adopt the innovation the information base available to potential adopters increases and the risk associated with the innovation decreases accordingly. The rate of diffusion increases but it cannot do so indefinitely. As the fraction of firms that have adopted the innovation increases, the number of potential adopters left decreases. Since the last potential adopters left are unlikely to be the most progressive firms the rate of diffusion decreases progressively until the process stops.

Such a diffusion process has a very simple mathematical representation. If $x(t)$ is the fraction of potential adopters who have already adopted at time t, then the rate of diffusion will be given by $\frac{dx(t)}{dt}$. Given the previous considerations this rate of diffusion can

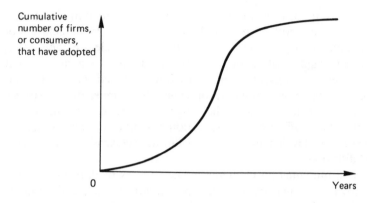

FIGURE 5.2   *Logistic diffusion curve*

be expected to be proportional both to the fraction of adopters $x(t)$ and to the fraction of potential adopters left $(1-x(t))$:

$$\frac{dx(t)}{dt} = \beta \, x(t). \, (1-x(t)) \tag{5.1}$$

where β is a constant. The differential (1) has the following solution:

$$x(t) = \frac{1}{1 + exp \, (-\alpha-\beta t)} \tag{5.2}$$

Equation (5.2) is the equation of a logistic time curve. This is a S shaped curve in which the rate of diffusion first increases until an inflexion point and then decreases (Figure 5.2).

In a typical diffusion study information is collected about the times at which each firm in an industry has adopted a particular innovation. On the basis of this information a graph of $x(t)$ vs. t is constructed. By fitting the logistic equation (5.2) to this data the best values of α and β are found. The meaning of α and β can be understood by reference to Figure 5.3. α determines the point at which the diffusion curve begins to rise and β the slope at which the curve rises. Therefore in Figure 5.3 curve 1 has higher values of both α and β. β is usually considered the diffusion rate constant. Values of α and β determined for different populations of firms can be compared and their differences interpreted on the basis of the different characteristics of these populations. For example, the relative technological progressiveness of firms in two different industries could be compared to the degree of industrial concen-

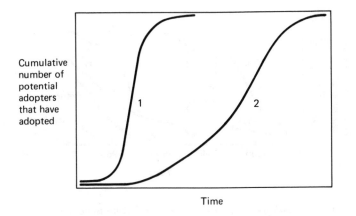

FIGURE 5.3 *Effects of an earlier start (α) and of a faster rate of diffusion (β) on the shape of a logistic curve*

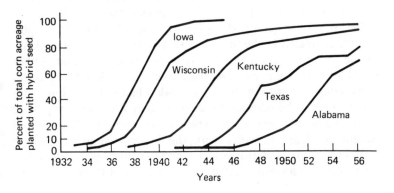

FIGURE 5.4 *Rate of diffusion of hybrid corn*

Source: Z. Grilliches, 'Hybrid Corn: An Exploration in the Economics of Technical change,' *Econometrica*, 1957, p. 502.

tration, to the skills employed and to the nature of the technology in the two industries. Likewise inter- or intra-firm diffusion may be studied.

The results of a number of studies based on the epidemic model can be summarised as follows. It was found that in general diffusion curves were S shaped and that a logistic curve would best explain the spread of innovations (Figures 5.4, 5.5), (Mansfield,

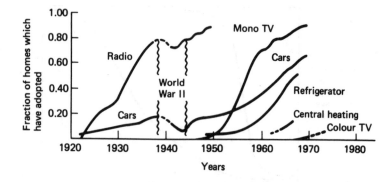

FIGURE 5.5   *Adoption of various items by UK households*

Source: I. Hendry. 'The Three Parameter Approach to Long Range Forecasting,' *Long Range Planning*, 1972.

1963a, 1969; Griliches, 1957; Norris and Vaizey, 1973; Nabseth, Ray, 1974; Metcalfe, 1970). The variables influencing the rate of diffusion were frequently divided into those related to the adopter and those related to the innovation. Among the variables related to the adopter, firm size was often the most important. Other variables in the same category were rate of growth of the industry and quality of management. On the other hand, innovations were usually seen to be adopted faster when they were relatively more profitable and less expensive. These conclusions are plausible and intuitively clear. Together with the intrinsic simplicity of the epidemics model and its relatively good fit with empirical data they go a long way towards explaining the success of the model.

Criticisms of the epidemic model and alternative models of diffusion

In spite of its undoubted success the epidemic model has lately come under considerable criticism. A number of authors have pointed out that the model is conceptually inadequate on a number of grounds. For example, the adopters' environment is assumed to be *homogeneous*, the only possible difference among adopters being some kind of progressiveness. Thus adopters were typically assumed to include a distribution of 'pioneers', 'early imitators' and 'laggards' (Rogers, 1962). The epidemic model does not consider the possibility that the rationality and the profitability of adopting a particular innovation might be different for different adopters; the adoption of a given innovation may be profitable for a particular firm at a certain time but may only become profitable for another firm operating in different circumstances at a later time. The two firms would be equally rational in adopting the innovation at different times and the real difference between them would be the conditions in which they are operating.

In addition to being homogeneous the adopters' environment in the epidemic model is also *static*: the population of the potential adopters and the diffusing innovations are assumed to be the same at the beginning and at the end of the diffusion period. On the other hand there is both theoretical and empirical evidence that many innovations undergo considerable changes during the course of diffusion and that these changes can both increase the number of potential adopters and, in turn, lead to subsequent modifications of the innovations themselves. The systematic trends in the evolution of technologies which were discussed in the previous section imply that the interaction of innovations with the adopter environment might show systematic changes during the maturation of a technology. Furthermore, innovations and technologies might differ with respect to the scope of the improvements which might be induced by the adopting environment.

Another important criticism of the epidemic model is that it takes into account only the adopters' or demand side of the diffusion process. However, if the diffusion is to proceed at all it must be profitable for both adopters and suppliers.

These inadequacies of the epidemic model have been addressed by several studies and a number of alternative models have been formulated, some of which will be reviewed here to provide examples of how they address the problems mentioned above. For more exhaustive surveys of diffusion studies, which do justice to all of them, the reader is referred to Stoneman (1983, 1984); David (1975); and Davies (1979).

Inter-firm differences are explicitly introduced in the models of David (1975) Davies (1979), and of Stoneman (1984). David's model is a probit model in which an innovation can be considered as a stimulus for a firm. Each firm has a critical level of stimulus and will adopt an innovation only when the stimulus represented by the innovation itself exceeds the critical level. The problem becomes therefore, to define the stimulus and the critical level. In David's model the critical level is defined by firm size. A new technique is assumed to have higher fixed costs but lower variable costs than an old technique per unit of output. If positive returns to scale are assumed, then at a given time, the adoption of the innovation will be profitable only for firms above a given size. Naturally both the evolution of the technology and the growth of firms can change both the critical level and the number of potential adopters. The time path of diffusion will be determined by the existing distribution of firm size, by the rate of growth of individual firms and by the evolution of capital and labour costs.

Inter-firm differences also play an important role in Davies' model. Due to uncertainty firms make decisions in a behavioural manner which can more appropriately be described as satisfising than as profit maximising. The most important aspect of inter-firm difference is the maximum acceptable pay-off period, R^*_{it}, which represents the equivalent of David's critical level. Firms are likely to differ not only with respect to the maximum period that they are prepared to wait before they obtain a pay-off from a given innovation, but in their estimate of the expected pay-off period, ER_{it}, for the innovation itself. Less knowledgeable firms are likely to either underestimate or overestimate ER_{it} and consequently to be either failures or late adopters. Adoption will take place when the expected pay-off period is less than the maximum acceptable pay-off period:

$$ER_{it} \leq R^*_{it} \qquad (5.3)$$

ER_{it} and R^*_{it} not only differ among firms at a given time but they are likely to change over time as knowledge of the new technology is accumulated in the industry. Firms' expectations of pay-off periods can become shorter and, if the profitability of an innovation is conclusively proved, firms might be prepared to wait longer before recouping the initial outlay (longer R^*_{it}). From these assumptions the probability can be derived that a firm of given characteristics will have adopted the innovation at a given time. Davies also assumes the probability of adoption to be proportional to firm size.

These features of Davies' model can be considered as innovator characteristics. However, a second important feature of Davies' model is the attempt that it makes to differentiate two types of innovation.

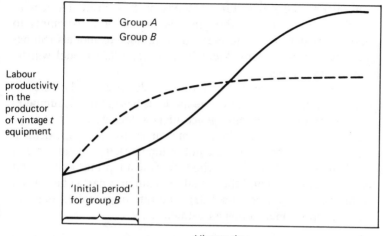

FIGURE 5.6   *Diffusion curves for group A and B innovations*

Source: S. Davies (1979).

The first type of innovation, called group A, consists of techno-logically simple and inexpensive processes which are built off-site. For this type of innovation learning effects are initially substantial but they soon fall away drastically, leaving the technology fairly stable at an early stage of the diffusion process. The second type of innovation, called group B, consists of technologically complex expensive innovations, usually built on a one-off basis and requir-ing lengthy periods of installation on the adopter's site. The 'lumpiness' and the small numbers in which these innovations are produced implies that learning effects will be initially slower than for type A. However, the scope for improvement will be greater for type B innovations and consequently, at a later stage, both the rate of diffusion and the ceiling for diffusion will overcome those of type A. Examples of type A innovations might be consumer goods (TV receiver, washing machine and so on) and of type B innova-tions a chemical plant or a steel mill. According to Davies the different types of innovation lead to different shapes of diffusion curves (Figure 5.6).

Another important deficiency of the epidemic model was the neglect of the supply side. An innovation can only be adopted if it is produced, and this in turn depends on how profitable it is to manufacture products embodying the innovation. Attempts to remedy this deficiency are made in a number of models (Stoneman, 1983, 1984). Here, Metcalfe's (1981, 1982) model will be used as an example.

In this model separate equations are derived for the rates of growth, supply and demand. The model is based on a Schumpeterian conception of the emergence of new technologies. The appearance of a new technology creates an *adjustment gap*, defined by the difference between the equilibrium market demand $n(p_n, \alpha)$ (the demand which will be reached after the adopting environment has been saturated) and the actual demand at a particular instant during the diffusion process $X_n(t)$. The rate of growth of demand $g_o(t)$ is proportional to this adjustment gap:

$$g_d(t) = b[n(p_n, \alpha) - X_n(t)] \tag{5.4}$$

An important feature of this model is that the equilibrium market demand $n(p_n, \alpha)$ depends on both the price p_n of the innovation and on the performance of the innovation which is here shown by α, representing the advantage of the new technology over the old one. Both product and process technology, leading to a decrease in p_n and to an increase in α, can therefore increase the equilibrium market demand. Post innovation improvements have explicitly been introduced into this model.

The rate of growth of supply $g_s(t)$ is assumed to be proportional to the profitability of the innovation for the supplier. This profitability will increase when the price of the innovation increases but will be negatively affected by increasing costs of all the inputs used in the production of the innovation (labour, materials, machinery and so on).

$$g_s(t) = \frac{p_n(t) - h_o - h_1 X_n(t)}{k} \tag{5.5}$$

where $p_n(t)$ is the price of the innovation, h_o, h_1 and k are constants, and k reflects the supply of capital funds and investment requirements. It is clear that the price of the innovation has opposite effects on supply and demand: a decreasing p_n will increase market penetration and the rate of growth of demand but it will influence negatively the rate of growth of supply. It is assumed that the rates of growth of demand and supply will tend to

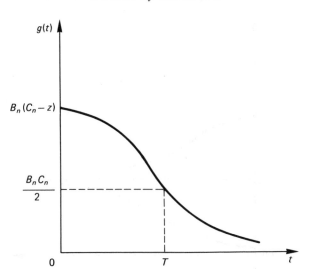

FIGURE 5.7 *Change of the adjustment gap during the diffusion of an innovation*

be equal at each point of the diffusion process, thus defining a balanced diffusion path. The resultant equation derived by equating those for the growth of demand and for the growth of supply is:

$$g(t) = B[C_n - X_n(t)] \tag{5.6}$$

where $g(t)$ is the rate of diffusion, B_n and C_n are parameters formally resembling the rate constant for diffusion β and the saturation constant K in the epidemic model, and $X_n(t)$ is the demand at time t. The parameters B_n and C_n are now jointly determined by the dynamics of demand and supply and they can change during the process of diffusion since they are influenced by the performance of the innovation.

As we have noted, Metcalfe's model is compatible with a Schumpeterian framework. In this model the temporary monopoly motivating an innovator to introduce an innovation into the market is represented by the adjustment gap $C_n - X_n(0)$, where $X_n(0)$ is the demand for the innovation at the beginning of the diffusion process. This adjustment gap will have its maximum value for the first innovator and it will gradually decrease as more imitators enter the market. A decreasing adjustment gap is reflected in a decreasing proportional rate of growth of demand (Figure 5.7).

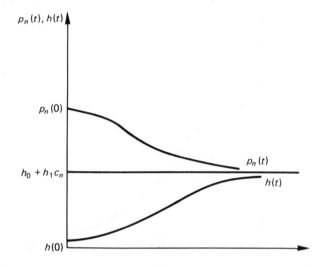

FIGURE 5.8   *Change in the demand for an innovation during its diffusion*

Simultaneously demand will increase along a logistic curve (Figure 5.8), innovation prices will decrease and production costs will increase (Figure 5.9) during the diffusion process. Consequently profitability for the innovator will decline together with the adjustment gap and the proportional rate of growth of demand. When the market is saturated the impulse created by the new innovation is exhausted.

The assumption that production costs are necessarily going to increase during the course of diffusion might seem questionable on both theoretical and empirical grounds. However, such a possibility is taken into account by the existence of post-innovation improvements. Post-innovation improvements can lead to decreasing innovation prices and to performance improvements. These are going to be reflected in an increasing scope for market penetration, or equivalently, in an increased value of C_n. Therefore the adjustment gap can increase during the course of diffusion (Figure 5.10). The same post-innovation improvements can lead to initial increases in the proportional rate of growth of demand (Figure 5.11) and in the profitability for the producers of the innovation (Figure 5.12). However, in the end diminishing returns

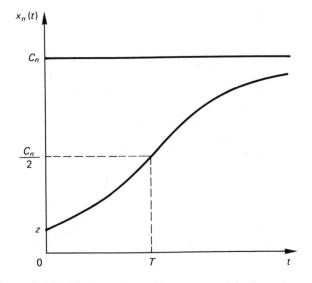

FIGURE 5.9   *Changes in the price ($p_n(t)$) and cost ($h(t)$) of an innovation during its diffusion*

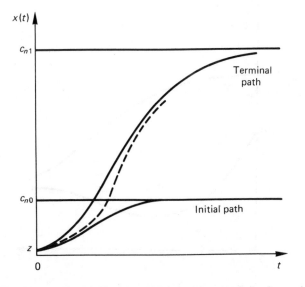

FIGURE 5.10   *Effects of post-innovation improvements in the evolution of demand for an innovation during its diffusion*

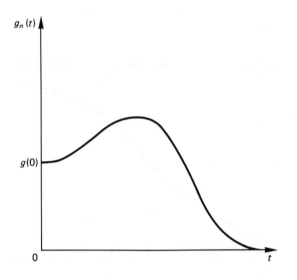

FIGURE 5.11 *Effect of post-innovation improvements on the adjustment gap during the diffusion of an innovation*

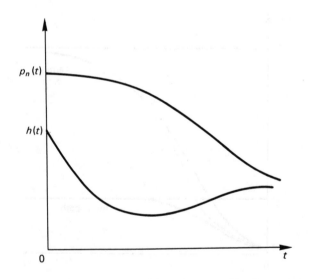

FIGURE 5.12 *Effects of post-innovation improvements on the price ($p_n(t)$) and cost ($h(t)$) of an innovation during diffusion*

to technology will set in and the scope for further post-innovation improvements will disappear.

The most significant aspect of the new models of diffusion which have been reported in this section is that they incorporate several of the more sophisticated elements of microeconomic thinking which have been prominent in other discussions of innovation. The possibility of inducement mechanisms, the effects of firm size, the existence of decision rules governing firm behaviour, and the interaction of Schumpeterian technical impulses with Schmooklerian demand influences are all present in these models. Furthermore, they include the evolutionary character of innovations which is captured in the notion of natural trajectories. The new models of diffusion therefore represent a genuinely qualitative step forward in the economic analysis of technological change. However, their novelty implies that they have not yet been subjected to the same degree of empirical testing as the older epidemic model. This is an important research priority for the future.

5.7 CONCLUSION

This chapter has been concerned with the existence of patterns in innovation and technical change. These patterns have been found both within the innovation process and within groups of related innovations by a large number of studies. From modest beginnings in classificatory and comparative studies, innovation research has progressed to formulate progressively more sophisticated microeconomic rationales for innovation which have had some success in explaining the data, but which are still in a process of evolution. Two important problems have been constant features of research in this field. The first of these is the analytical relationship between technological and demand factors, expressed in the contrast between the Schumpeterian and Schmooklerian perspectives. The work discussed above has made some important contributions to resolving that contrast in a fruitful way. The second problem relates to the method of aggregation used in looking at patterns in large numbers of innovations. There is a tension in the literature between aggregation to the level of the market, and aggregation to the level of 'technology type'. This is particularly clear in the literature on market structure and innovation, but pervades other topics also. This issue is important because the aggregation to market or technology is the intermediate step in the movement toward macroeconomic aggregation. These issues are therefore further discussed in the next two chapters, which deal with this level of aggregation.

6 Technological Change, Output, Employment and Trade

The discussion in Chapter 5 dealt mainly with the factors contributing to the origin of technological innovation. However, the real importance of innovation lies in its potential to generate greater wealth and better living conditions. Innovation can have these effects only after it has been adopted by some firms and institutions, and its overall effect on a given economy will depend on its degree of diffusion. In order to understand these effects of innovation the analysis in this chapter will move from the predominantly microeconomic level of analysis of Chapter 5 to higher levels of aggregation and will give more specific consideration to time patterns of economic development which are related to technological change.

Many of the changes which have taken place in the last two hundred years in industrialised societies are, if not caused by, at least related to technological change. One of the most important differences between now and then is the variety of new products and techniques which are used in everyday life in households, firms and in public life. Not only have all these new products come into existence but many of them, after initially being reserved to the upper echelons of society, have become gradually cheaper and therefore available to a majority in the industrialised countries. In the course of time the price of these products has declined as a proportion of average income in such a way that many citizens can now afford them. The cheapening of products has occurred by means of an increasing efficiency in the processes by which they are produced. Thus two of the consequences of technological change are the continuous emergence of new products and the continuous cheapening of existing products. These changes are those that directly affect the users of products. The cheapening of products however, is obtained by increasing the efficiency of the processes by which they are produced, which in turn leads to a reduction in the inputs required to produce one unit of output. Most industrial processes use labour and consequently technological change has very important implications for employment.

By taking a point of view different from that of the user of products but observing events at a higher level of aggregation one can see that the two previously mentioned changes have been accompanied by changes in the composition of employment and output. As new products emerged and the manufacture of old products became more efficient, old products came to account for a declining proportion of employment and output. Quite apart from their effect on total employment these changes required different types of skills. In the course of time therefore, the economies of most countries have undergone both quantitative and qualitative changes. Not only is more available for the average citizen but more of new types of goods and services. In many cases this may only be true for the mythical 'average' citizen and, at least during certain periods of time, economic growth may be accompanied by increasing income inequality.

The changes which have so far been described relate mostly to the national level but very important consequences of technological change have taken place at the international level as well. For example, international trade is influenced by patterns of specialisation and of technological leadership. Some types of goods are produced for the first time in one or a small number of countries from which they are initially exported. For a period of time trade in these goods flows from the innovating countries to other countries which cannot produce them but which are potential adopters.

These examples serve simply to illustrate the range of economic phenomena in which technological change plays a fundamental role. In general it can be said that the world economic system undergoes a process of expansion and transformation in which the quantity of goods and services available to the 'average citizen' is continuously increasing, and also in which the types of goods which are produced, the skills used, the location of production and the patterns of trade are continuously changing.

However, the concept of economic growth which has been used in a number of economic theories is narrower and more specialised than the processes just described. This narrower concept could be summarised as follows: technological change, seen as the generation of a stream of innovations and their diffusion, leads to increasing productivity or, equivalently, to a movement of the isoquants of the production function towards the origin; unit costs fall; and a greater quantity of goods and services can be produced by a constant population. Population growth and improvements in organisational efficiency can lead to further economic growth. Clearly this concept of economic growth and the related concept of productivity capture only one of the dimensions of the process of economic development previously described, namely the cost-decreasing dimension. Productivity grows over time when, by using a constant set of inputs but by changing techniques, a greater quantity of a qualitatively constant output can be produced. However, the concept of productivity is unproblematic only in this highly simplified situation. If the nature of the inputs used and of the outputs produced change over time or if, at a higher level of aggregation, the composition of the national output changes substantially, to define and measure productivity becomes far more complex (Rosseger, 1980). Other factors such as firm organisation, educational levels of the labour force, union strength, government intervention, capacity utilisation or patterns of international trade can contribute to increase economic efficiency and to complicate the concept of productivity.

In spite of these difficulties productivity growth patterns for many economies have been studied. Sometimes the productivity of one factor of production is studied, for example of labour, by measuring output per worker per hour. Alternatively the effect on productivity growth of all factors may be studied by measuring total factor productivity (Rosseger, 1980). To measure total factor productivity all the factors which influence productivity should be known. Generally many potential causes of productivity growth cannot be identified or measured (Nelson, 1981) and a compromise has to be reached between comprehensive and analytically satisfactory models. This tension and the simplifying assumptions used are discussed in the next section in which some theories of economic growth and the role played in them by technological change will be described.

6.1   CLASSICAL ECONOMISTS AND ECONOMIC GROWTH

The importance attached to the problem of economic growth has varied considerably during the history of economic thought. Classical economists were very concerned with it. Adam Smith in *The Wealth of Nations* (1776) identified the division of labour, free markets, and technical change in the form of new machines as the three important causes of increasing incomes. Malthus and Ricardo thought that the process of economic growth would come to a halt due to the limited quantity of land available (Kindleberger, 1962). According to Malthus, when the existing land had been saturated food shortages and starvation would limit further growth of the population. According to Ricardo the most productive areas of land would be occupied first and the quality of the land to be occupied subsequently would gradually decline. As a consequence diminishing marginal productivity would be encountered in expanding the land under cultivation until no further increase in output could be obtained (Kindleberger, 1962). It is interesting that especially in the case of Ricardo, who was quite aware of the importance of technological change in manufacturing industry, no such contribution to raising agricultural productivity was considered possible (Hertje, 1977). Marx argued (1887) that the growth of the capitalist system would eventually come to a halt. A crisis in capitalist society would result from the inevitable decline in the rate of profit, which would be brought about by production becoming more and more capital-intensive. The increasing ratio of constant capital (capital equipment, buildings, etc.) to variable capital (wages) $\frac{c}{v}$, which Marx called the organic composition of capital, would lead to a fall in the rate of profit since a surplus can only be extracted from labour power. It is not impossible for other tendencies to counteract the falling rate of profit. For example an increase in the rate of surplus could compensate for the increasing organic composition of capital. There is still disagreement among marxist economists about the extent and composition of such compensation (E. Mandel, 1969). The use of machines was therefore both a strength and a weakness of the capitalist system, a strength because it allowed vast increases in productivity, a weakness because in the long term it could decisively contribute to the collapse of the system itself.

Thus the problem of economic growth occupied a central place in the thinking of classical economists although it was analysed in ways which are very different from those prevalent nowadays. Furthermore, technical change, though not thoroughly formalised, played an important part in their thinking. But economic growth ceased to be considered an important problem after the emergence of the neoclassical school in the second half of the nineteenth century. It regained an important position in economic thought only after the Second World War.

6.2 THE HARROD–DOMAR MODEL

One of the first theories of economic growth which was developed after the Second World War was due to Harrod (1948) and Domar (1946). The Harrod–Domar model was conceived as a result of the Keynesian revolution in income theory and of its preoccupation with the correct balance between savings and investment. Keynes pointed out that income would be stabilised, even if possibly at less than full employment, when new expenditure for capital equipment coming into the system exactly offset savings taken out of the system. However, this is true only for the short run. Harrod pointed out that investment in period t would create additional capacity in a subsequent period $t+1$ and that demand based on the income of period t would not necessarily be enough to absorb the greater output due to this extra capacity. In other words, a correct balance between savings and investment has to be maintained in the course of time to ensure a 'balanced' economic growth. Naturally the extra capacity varies according to how 'productive' the new capital equipment is. This productivity is measured by the capital/output ratio $k(k = \frac{K}{Q} = $ quantity of capital required to produce one unit of output) of the new capital equipment. The lower the capital output ratio the greater will be the extra capacity generated by a given volume of investment. Starting from these premises it can easily be proved that the growth rate G is related to the savings ratio s and to the capital output ratio k by the following equation (Kindleberger, 1962):

$$G = \frac{s}{k} \tag{6.1}$$

where:

$$s = \frac{S_t}{Y_t} = \text{fraction of income saved} = \frac{I_t}{Y_t}$$

S_t = savings in period t

I_t = investment in period t.

According to equation (6.1) growth rates can be increased either by increasing the savings ratio or by lowering the capital/output ratio. The Harrod–Domar model exerted a considerable impact on economic planning in developing countries although it had been conceived mainly on the basis of an analysis of industrialised economies (Colman, Nixson, 1978). This is not the place to discuss the general merits and limitations of the model but some comments about the role played in it by technological change are relevant here. First, it can be observed that in equation (6.1) the rate of economic growth is independent of labour and that consequently the Harrod–Domar model relies largely on a capital theory of value. As a consequence, labour can be introduced into a Harrod–Domar system only at a constant capital/output ratio, thus largely ruling out the possibility of mutual substitution between capital and labour. Secondly, technical change does not enter directly in the Harrod–Domar model. Technical change can influence the capital/output ratio k, but in equation (6.1) k is considered either a constant with respect to time or variable in unknown ways and therefore a datum for the theory.

The shortcomings of the Harrod–Domar approach led to the formulation of neoclassical models of economic growth which allowed changes in labour as well as in capital and the mutual substitution of labour and capital.

6.3 NEO-CLASSICAL MODELS OF ECONOMIC GROWTH

In neo-classical models of economic growth a production function analogous to that defined for a firm in Chapter 2 can be defined for the whole economy:

$$Q = f(L, K) \qquad (6.2)$$

This production function fixes the maximum output Q which can be obtained with given levels of inputs, L labour and K capital. Economic growth will then take place either through increase in inputs along a given production function or through a shift to a more efficient production function. Technological change can be expected to shift the production function towards the origin, in the direction of a more efficient utilisation of inputs. The problem then arises of analytically separating the relative contributions of input increases, and of shifts in the production function, to the actual productivity growth and economic growth that have taken place.

The same assumptions which applied to the microeconomic production function (p. 24) apply at the macroeconomic level. So, for example, firms and therefore the whole economy will tend to maximise profits, given demand and factor prices. Also, firms will be price takers, or equivalently, no monopoly elements will be present in the system. Information will be freely available or technological knowledge will be a public good. Supply and demand will generally be in perfect equilibrium. When the contributions of different factors have to be combined the weights used will be equal to the marginal products of the factors or, equivalently in neoclassical economics, to their prices (e.g. wages and rates of interest). However, a number of assumptions which were not required at the microeconomic level are required at the macro level. For example it is commonly assumed that there are constant returns to scale, or that the output per unit of inputs is independent of the total quantity of inputs, and that technical change is neutral. The importance and limitations of these assumptions can be illustrated by reference to one of the best known neoclassical studies of economic growth, that of R. M. Solow.

The aim of Solow's study was to find 'an elementary way of segregating variations in output per head (labour productivity) due to technical change from those due to changes in the availability of capital per head' (Solow, 1957). In other words, what have been the relative contributions of technical change and of the larger quantity of machines per employed worker to the observed pattern of productivity growth? Notice that here productivity means labour productivity and not total factor productivity.

The aggregate production function can be written as:

$$Q = F(K, L, t) \qquad (6.3)$$

where K, L and t are capital, labour and time respectively. The variable t here allows for technical change. In other words for Solow technical change is 'a shorthand for any kind of shift in the production function'.

If technical change is *neutral* (see p. 26) the production function can be separated into two components, one $A(t)$ dependent only on technical change and the other one $f(K, L)$ dependent only on the inputs of labour and capital:

$$Q = A(t)f(K, L) \qquad (6.4)$$

If technical change were not neutral the relationship between inputs (K and L) and outputs (Q) would change with time and, therefore the production function could not be separated into two components.

Furthermore, if technical change shows constant returns to scale (i.e. if output per unit of capital equipment does not depend on the size if the producing units) Q and K can be replaced by $\frac{Q}{L} = q$ (labour productivity) and by $\frac{K}{L} = k$ (capital/ labour ratio). With these two assumptions (6.4) can be transformed into:

$$\frac{\dot{q}}{q} = \frac{\dot{A}}{A} + w_k \frac{\dot{k}}{k} \qquad (6.5)$$

where the dots indicate time derivatives and w_k is the share of capital in total output. From (6.5) the technical change factor A can be empirically estimated if time series of output per man hour, capital per man hour, and the share of capital are known.

This theoretical framework was applied by Solow to the USA economy for the period 1909–1949. During this forty-year period output per worker per hour approximately doubled while at the same time the cumulative upward shift in the production function was about eighty per cent. Consequently, according to Solow, about one-eighth of the total increase was traceable to increased capital per man hour and the remaining seven-eighths to technical change.

To the extent that one can trust these results technical change emerges as by far the most important source of economic growth. But what technical change is it? Solow's definition, given at the outset of his paper equates technical change to anything that causes a shift in the production function. Consequently the A factor does not include contributions from increased inputs of capital and labour but it could include anything else such as, for example, better organisation, improved technological knowledge, or better education of the labour force. Therefore, the *A* factor has been called a *residual*, or what is left of economic growth after one has accounted for the effects of increased inputs of labour and capital. This residual character of technical progress is not an unexpected outcome of Solow's neoclassical approach, but a necessary consequence of defining technical change by exclusion rather than modelling it explicitly. These criticisms are not meant to deny the value of Solow's paper, which has rightly remained famous as one of the most important examples of the neoclassical approach to economic growth. Solow, by showing how quantitatively limited is the contribution of increased labour and capital inputs has demonstrated the need for a better understanding of what is hidden in the residual *A*.

Two of the most important assumptions in Solow's paper were those of neutral technical change and of constant returns to scale. Solow's tests showed that these asssumptions are plausible at the macroeconomic level. However, we know that increasing returns to scale have been important in the development of some industries and that there are well documented examples of biased technical change. The violation of these assumptions at a lower level of aggregation is not necessarily incompatible with their validity at the macroeconomic level, but the interesting question here is what mechanisms bring this about. No answer can be found in the neoclassical literature in which aggregation is implicitly assumed to be additive or in which, in other words, total output of the economy is the sum of the outputs of qualitatively identical units without any interaction terms.

A further important assumption contained in Solow's approach is that technical change is considered to be disembodied, or independent of capital and labour. Thus the production function of the whole economy may shift inwards and the same quantities and types of capital and labour will then produce more output. If this assumption were true we could, for example, be using the same machinery and quality of labour as at the beginning of the industrial revolution but producing a much larger output due to improvements in organisation, knowledge, and so on.

Clearly, this assumption is not entirely realistic. Certainly the machinery presently in use is generally much more efficient than that used at the beginning of the industrial revolution and the labour force more educated and skilled. Therefore, at least part of the technical change which has taken place has been *embodied* in machines and labour of higher quality. The concept of different *vintages* of labour and capital has been proposed to describe these qualitative changes: more recent vintages would have a higher efficiency and through their adoption the average productivity of the economy would gradually increase.

The difference between the embodied and disembodied assumptions can be compared to two different accounting methods: the contribution of increased quality of capital equipment would be contained in the residual *A* following the disembodied assumption whilst it would be included in the contribution of capital following the embodied assumption. Consequently passing from the disembodied to the embodied assumption would decrease the value of the residual *A*.

A study attempting to separate the qualitative and quantitative contributions of the factors which influenced US economic growth between 1929 and 1957 was that of Denison (1962). Denison identified a long list of sources of growth (for example labour, capital, education, hours of employment, changes in the age–sex composition of the labour force, foreign assets and economies of scale) and tried to adjust them for their change in quality during the period under study. At the conclusion of his study Denison found that there had been five main sources contributing to US economic growth between 1929 and 1957. These were: increased education (23 per cent), increased employment (34 per cent), increased capital input (15 per cent), the advance of knowledge (20 per cent), and economies of scale (9 per cent). Here also the contribution of the advance of knowledge was obtained as a residual. However, by accounting explicitly for a larger number of sources of growth the size of the residual had decreased. The clear implication that can be derived from a comparison of Solow's and Denison's studies is that by accounting explicitly for more and more sources of growth the size of the residual would decrease until eventually it contained only measurement errors. This justifies the fact that the residual has been called by some scholars a 'measure of our ignorance'. In the context of the present book it must be emphasised that our ignorance seems to be greater about technological change than about any of the other potential sources of economic growth, in spite of the great importance generally associated with technological change as a source of long-term economic growth. One of the problems of the previous studies, in addition to the neglect of some important variables which will be analysed in the next section, consists in measuring technological change *implicitly* by means of other variables which are influenced by technological change. In order to reach a better understanding of the economic effects of technological change we need the opposite approach: modelling technological change explicitly on the basis of studies of the evolution of technologies.

6.4 HETERODOX APPROACHES

Neoclassical models of productivity and economic growth achieved a great analytical clarity by concentrating on a limited number of variables. Consequently within this approach all the variables which were not included contributed to the residual factor. However, that many other variables are potential co-determinants of productivity growth has been shown by a large number of studies, sometimes originating from different research traditions (for example, organisation studies) and generally less rigidly based on a theoretical model than the previous ones.

Following Nelson (1981), the extra factors that these studies have shown to be potential co-determinants of productivity growth can be divided into three classes:

1. Co-determinants of productivity growth at the level of the firm.
2. Variables related to the nature of technological advance.
3. Mechanisms of interaction between the primary determinants of productivity growth.

These can be considered in turn.

The neoclassical view of the firm (Chapter 2) was compatible with that of those organisational theorists who conceived organisations as machines with human elements. More recently organisational theorists have considered organisations as social systems in which multiple groups with conflicting interests coexist. Within this perspective, problems of motivation, coordination and intra-institutional conflict acquire a much greater importance. Consequently variables such as workers' satisfaction, degree of unionisation and quality of management are seen to be important co-determinants of productivity growth.

Secondly, as Part I of this book has sought to demonstrate, technological change is a process substantially different from the routine repetition of established activities. For example, R & D is a highly uncertain process. Many institutions undertake similar R & D projects. The likelihood that each R & D project is successful is in this way increased, but whether the chosen projects are socially optimal depends on the existing regime of property rights (for example, patents). On the other hand, sometimes learning by doing is a source of new technology as important as, or more important than R & D. Furthermore, imitation lags in the diffusion of technology and differences in productivity among firms in the same industry and in the same country show that the process of acquisition of information, rather than being costless and requiring no time, is one of the fundamental resource-consuming aspects of technological change.

In growth accounting studies a number of different sources of growth are identified, and their contributions estimated separately. Quite apart from the problem of adding up the various contributions in this way, the mechanisms of interaction between the primary sources of growth are not addressed. Examples of this kind of interaction are those between education, R & D and capital equipment, or those between the labour process and the design of technology.

According to Nelson those studies which have highlighted the importance of previously-neglected sources of productivity growth have cast doubts on the extent to which the problem of productivity growth is amenable to neoclassical analysis. However, these more heterodox studies have not yet provided stable patterns or relationships between variables which could be used for modelling purposes. A step in this direction has been taken by Nelson and Winter with their evolutionary model of economic growth.

6.5   AN EVOLUTIONARY APPROACH TO ECONOMIC GROWTH

Nelson and Winter's (1974) analysis of economic growth is both Schumpeterian and centred on a behavioural theory of the firm. It is thus closer to our analysis in previous chapters and departs from neoclassical approaches. The environment within which firms operate is characterised by uncertainty, struggle and motion rather than by careful calculations over well-defined choice sets. It is not equilibrium, but the forces tending to create change and therefore to destroy existing equilibria which are the main driving forces in the system. The firms which operate in this dynamic selection environment are not profit maximizers but satisficers. Firms are all different and the state of each individual firm at a given time is characterised by its production technique – described by a pair of input coefficients (a_L, a_K), which are the quantities of labour and capital required to produce one unit of output – and its capital stock K. The state of an industry at a time t is the collection of the states of those firms which constitute the industry. Individual firms have targets and decision rules (Chapter 5) and there is a considerable variance amongst firms in factor ratios, efficiency and rates of return. Firms follow existing decision rules until targets are exceeded and only when they are not will search processes lead to changes in existing decision rules. The search for new decision rules is *local* in the sense that firms first try to explore techniques and methods which are similar to those that they or other firms in the industry are presently using. The general rate of technical change is determined by how local is this search. Technically advanced firms reinvest their profits and expand. In this way they both set a precedent for other less progressive firms and drive up the wage rates facing other firms. Less progressive firms are therefore stimulated to imitate the new techniques both because they are more efficient and because the higher wage rates would make the old techniques relatively even less efficient. As a result of this competitive process industry states and therefore the state of the economy change over time.

A number of other assumptions concerning financing, devaluation and employment, which cannot be discussed here for reasons of space, are necessary to define the model. Those assumptions which have been discussed – the non-optimising nature of firms, the local character of search and imitation, the influence of individual firms' decisions on the selection environment – differentiate clearly the Nelson and Winter evolutionary world from that of neoclassical economics.

Nelson and Winter incorporate these assumptions in a computer simulation of the behaviour of the US economy in the period

1909–49. In 1909 their firms were given values of input coefficients (that is, their techniques) derived from Solow's historical time series. Also values of the depreciation rate, the critical rate of return and the rate at which the labour supply curve shifts to the right were based on the same time series. Using these chosen values and based on the mechanism previously described the model generates its own time series of Q/I (output/labour ratio), K/I (capital/labour ratio), wage rates, capital share and of Solow's technology index A (the residual). These simulated time series compare well with Solow's historical time series except for the periods of the two world wars.

The results obtained with this evolutionary model of economic growth are therefore quite interesting. The model starts from microeconomic assumptions considerably different from those of neoclassical economics yet a macroeconomic world similar to that used by Solow in his neoclassical analysis of economic growth is produced. Thus Nelson and Winter argue that their model proves that what can at an aggregate level be analysed as a neoclassical world, is at a microeconomic level an evolutionary world. Their model contains assumptions which are certainly more realistic than those of neoclassical economics although in its present form they are still highly simplified. For example, a firm's technique is still represented exclusively by its capital and labour coefficients. However, what is interesting for students of technological change is the possibility that this model offers to model firm behaviour, including the firm's approach to technological change, on the basis of empirical observations. It would be particularly important to be able to incorporate into such a model concepts derived from studies of innovation and of technological change such as natural trajectories, technological opportunity, and other topics discussed in Part I of this book.

Nelson and Winter's model then shows that non-neoclassical firms and mechanisms of competition can simulate the process of economic growth of a real economy. Although this is an important result which incorporates some of the important factors identified by Nelson (1981) as co-determinants of productivity growth and as special features of technological change, it is still limited in its approach to problems of aggregation, interaction effects, and industrial structure. These latter issues are given more prominence in the neo-Ricardian approach of Pasinetti (1981).

6.6 PASINETTI'S MODEL OF ECONOMIC GROWTH AND STRUCTURAL CHANGE

Relationships between industrial sectors are a central feature of Pasinetti's model of economic growth and structural change. The economy is considered as the sum of a series of 'vertically integrated sectors'. All the materials and intermediate inputs required to produce the final outputs of each sector are produced in the sector itself. Production takes place by means of labour and capital goods. The household or final sector provides labour to the productive sectors and exerts the demand for final goods, which can be either consumption or capital goods. A series of equations describe the flows into and out of each vertically integrated sector. The system is in equilibrium when there is full utilisation of the existing labour and productive capacity. The equation expressing this equilibrium condition contains the sum of the contributions of each vertically integrated sector. In turn each sectoral contribution depends on both technical efficiency of the sector and on the demand for its outputs. For example, a sudden increase in technical efficiency without a corresponding increase in demand could lead to labour displacement in the particular sector considered. However, equilibrium for the whole system can be achieved even when in some of the vertically integrated sectors demand and technical efficiency are out of phase, since for example an excess of labour in one sector could be compensated by a shortage in another sector. Therefore the particular type of aggregation which is implied in Pasinetti's economic system is not based on the concept of an 'average' vertically integrated sector, which is a microeconomic analogue of the whole economy, as it was for example in the Harrod–Domar and neoclassical theories of growth.

Given an economic system with the basic characteristics described above, Pasinetti outlines three different scenarios under which the expansion and transformation of the system can take place. In the first scenario economic growth takes place by means of population growth but without technical change. In this scenario the productive capacity for each commodity must increase just enough to be able to satisfy its demand. This implies that in order to endow each sector with equilibrium productive capacity new investment will have to be equal to final demand. This leads to a Harrod–Domar equation for each industrial sector. In these conditions each vertically integrated sector has exactly the amount of productive capacity required to maintain full employment. The Harrod–Domar equation emerges from this model as describing the subset of Pasinetti's economic system in which the expansion of the system takes place by population growth alone without technical change.

In the second scenario the economy grows because in addition to population growth there is also technical change, although it is technical change of a particular type. Technical change is equivalent to productivity growth and uniform technical change means that all vertically integrated sectors have the same rate of productivity growth. All sectors will therefore be able to increase their productive capacity to the same extent. In order for the economy to grow by maintaining full employment and full utilisation of its productive capacity the increased productive capacity due to productivity growth must be matched by a demand which increases at the same rate. In general the rate of growth of demand for various commodities varies depending, among other parameters on their income elasticity of demand. However, in this second scenario the expansion of demand is assumed to be uniform, which means that consumer tastes are invariant with respect to income. In this case, with uniform technical change and uniform expansion of demand, all sectors grow at the same rate which is the sum of the rates of population growth and of technical change. Here as well as in the first scenario one can think of an average sector which constitutes a microeconomic analogue of the whole economy and as a consequence as change in the composition of output and employment will take place during growth.

In the third scenario rates of productivity growth and the expansion of demand are allowed to differ amongst different vertically integrated sectors. According to Pasinetti technical change can be expected to be uniform through time while the income elasticity of demand will decline for some products, for example for those that are near market saturation. In this case, if the expansion of demand slows down relative to productivity growth, employment growth will slow down in the respective sectors even leading to an absolute reduction in employment. In this more general and more realistic case the composition of the economic system will continuously change, both in terms of output and of employment, and this change is required if the process of economic growth is to continue. In this third scenario the concept of an 'average' sector loses a great part of its validity. The whole economic system can grow with full utilisation of labour and productive capacity even when particular vertically integrated sectors lose employment and productive capacity. Naturally for this to occur different vertically integrated sectors must be 'synchronised' in such a way that employment is transferred from the sectors shedding it to those which are having particularly high rates of employment growth and consumer income must be reallocated in line with the new structure of output. Clearly the nature of aggregation in such an economic system is not purely additive and it would be very difficult to describe the economy as the sum of 'average' vertically integrated sectors neglecting the processes of interaction amongst them.

Thus this model, while not having the microeconomic richness of Nelson and Winter's model, has important extra dimensions of realism in its account of the range of commodities and inter-industry differences in technical change and demand. These issues are important in our analysis of structural change effects on long-term economic growth in Chapter 7. They also lead us naturally to the issue of technical change and employment.

6.7 TECHNOLOGICAL CHANGE AND UNEMPLOYMENT

The classical economists such as Ricardo, Malthus, Stuart Mill and Marx were particularly concerned with the problem of employment implications of technical change. Their ideas can be summarised by means of the so-called 'compensation theory' (Heertje, 1977). The introduction of machines, by making production more efficient can lead to the displacement of workers. However, these workers will not necessarily remain unemployed because by means of a variety of compensating mechanisms they can be reemployed either with the same occupation or in a different branch of production. Any resulting unemployment will then be due to the balance between displacement and compensation. Classical economists differed about the extent of compensation which they considered possible.

Among the possible compensating mechanisms some are

endogenous in that they do not require further capital formation in addition to the new machines. For example, the greater demand created by the cheaper goods produced with the new machines could lead to the reemployment of the displaced workers making the same goods as before. Clearly, whether this mechanism can lead to full compensation depends on the extent of wage flexibility and on the demand for goods. Both wage inflexibility and a failure of demand to increase in line with the greater availability of goods (for example due to a low price elasticity of demand) would limit the possibility and extent of compensation.

Other possible compensating mechanisms are exogenous, which means that they operate independently of the application of the new techniques. An example of this form of compensation occurs when displaced workers are reemployed in the production of the new machines.

Doubts existed amongst classical economists about the possibility that this mechanism could lead to full compensation. Even if the number of vacancies in the production of new machines were equal to the number of jobs displaced in the industries in which the new machines were applied, the displaced workers might not have the right skills. In this case structural unemployment would result. Naturally in the long term retraining could lead to full compensation but this would not prevent some form of temporary disequilibrium.

In the production function approach some of the problems of compensation theory can be seen in a different light. For example the substitutability of labour and capital determines the extent of compensation which is possible. If only particular combinations of labour and capital can be used the possibility of compensation will be more limited than in the case of complete substitutability. Also, a biased technological change tending to save labour preferentially can limit the extent of compensation with respect to neutral technological change (Stoneman, 1983).

A somewhat different point of view can be found in Pasinetti (1981). If technical change and demand do not grow at the same pace in each vertically integrated sector then overcapacity and consequently unemployment can arise in at least some sectors. In time unemployment at the level of the system can be avoided only by transferring some labour to vertically integrated sectors characterised by a higher rate of growth of demand. Endogenous compensation seems therefore impossible in Pasinetti's approach. Compensation is possible but according to Pasinetti it will not necessarily be complete even when labour is moved from slower growing to faster growing sectors. There may be situations in which full employment in presence of different rates of technical change and of growth of demand in various vertically integrated sectors will only be achieved by means of government policies specifically aimed towards that goal (Pasinetti, 1981). Pasinetti's approach differs from those previously described in that structural change is explicitly incorporated into his economic system. The presence of structural change implies that a static version of compensation theory cannot account for observed patterns of employment. This is a very important theme which will be discussed more fully in the next chapter.

To summarise, unemployment arises from the imbalance between job displacement due to the introduction of new techniques and compensation. The extent of compensation which can take place depends on the type of technical change (neutral or biased), on wage and price flexibility, on the rate of diffusion of the new techniques, and so on. Compensation was previously discussed implicitly or explicitly in terms of an average sector, implying that if compensation was possible for this sector it was equally possible for the whole economy. More recent theories and empirical evidence seem to imply that compensation, if possible, necessarily implies a changing composition of the output and of the labour force amongst different industrial sectors.

6.8 TECHNOLOGICAL CHANGE AND INTERNATIONAL TRADE

The previous discussion of the effects of technological change on output and employment was implicitly concerned with the level of aggregation of a national economy. However, a large proportion of the goods produced in each country are exported. International trade is itself affected by technological change and in turn influences economic growth and employment. Flows of trade depend on the countries in which particular goods and services are produced and on those where demand for the same goods and services exists. These patterns of production and use are amongst the main problems that a theory of international trade has to explain. Moreover an important change has taken place during the twentieth century in the way in which goods and services are traded between different countries. This change has been represented by the emergence of the multinational enterprise, MNE (otherwise called the multinational corporation, MNC, or the transnational corporation, TNC). Trade between independent firms (inter-firm trade or arms length transactions) has been partly replaced in the international arena by trade between subsidiaries of the same MNE (intra-firm trade). This change has taken place gradually but its pace has accelerated since the Second World War. The existence of MNEs has created further problems for theories of international trade. In addition to having to account for the type, directions and intensities of flows of trade these theories now have to explain why intra-firm trade has in many cases replaced inter-firm trade and what effect this change has had on the previously mentioned flows. In what follows some theories on international trade will briefly be discussed stressing the role played within them by technological change. The problem of the directions of trade flows will be analysed first, passing subsequently to that presented by MNEs.

Traditional theories of international trade relied on the concept of *comparative advantage*. Countries are differently endowed with natural resources, labour, skills, capital and so on. As a consequence of these different endowments some countries can be expected to have a comparative advantage in the production of certain goods. For example, the combination of quantity and quality of land and of weather conditions might give some countries a comparative advantage in the production of wheat and other countries a comparative advantage in the production of coffee. More generally a country could be particularly well endowed with a number of factors of production which would give the country a comparative advantage in the production of those goods in which the factors of production are important inputs. In this context comparative advantage refers to the *relative* efficiency of production in the same country: if a country produces textiles *relatively* more efficiently than food that country will benefit from exporting textiles and importing food even with countries which are absolutely more efficient in the production of both. Total output is increased by trade. Specialisation in the products in which they have a comparative advantage is the prescription for national economies following from the comparative advantage theories of international trade.

The comparative advantage theory of international trade gained prominence during the nineteenth century and has remained the most widely accepted theory until the present time. However, especially after the Second World War, considerable empirical evidence about patterns of trade has been gathered which fits rather uneasily within the comparative advantage theory. For example in the early 1950s one would have expected that the US, being the country with the most expensive labour and the most capital intensive economy in the world, should have exported capital intensive goods and imported labour intensive goods. However, as it turned out the US was exporting goods which were more labour intensive than its exports. This paradox for the comparative advantage theory was called the 'Leontieff paradox'. (Leontieff, 1953).

The Leontieff paradox led to the emergence of the so-called 'new technology' theories of international trade (Posner, 1961; Vernon, 1966; Wells, 1972; Soete, 1982). One of the first and perhaps the best known of these theories is the 'product cycle' (Vernon, 1966). According to this theory entirely new products are more likely to originate in high income countries, where both the demand for them and the skills required to produce them exist. For a period of time after the emergence of a new product, manufacture will be geographically limited to the country of origin. Demand for the same product in other high income countries will during this period be satisfied by means of exports. When a sufficiently large market for the new good has been created in other high income countries manufacture of the product will begin in these other high income countries. The flow of international trade from the originating to the other high income countries will gradually decrease. In the end the direction of the flow of trade may be reversed if the imitator countries have lower costs and become relatively efficient in the production of the new good. As the product matures it becomes more standardized and its production is easier to transfer to other countries of gradually lower income and skills. In the end it is possible for most of the world trade in this once new product to flow from low wage to high wage countries.

Vernon (1979) has recently re-examined the product cycle hypothesis in the light of the growing internationalisation of multinational enterprises and of the decreasing income gap between the US and the rest of the world. It is now possible for the introduction of a new product to be conceived and planned as a global operation from its very outset. However, there are still a large number of situations (for example a US firm, or an LDC-based firm setting up their first foreign subsidiaries) to which the initial hypothesis would apply.

The product cycle is closely related to the Schumpeterian theory of economic development. The country which first manufactures a new product enjoys a temporary monopoly in its manufacture. During this period it can charge monopolistic prices for its exports. As the manufacture of the new product diffuses internationally its price will gradually decrease due to the elimination of supranormal profits.

The product cycle proved very useful in explaining the Leontieff paradox. The manufacture of a particular good is likely to become less labour intensive as the good 'matures' and therefore to have the maximum labour intensity for a completely new product. The Leontieff paradox could be explained if US exports were mainly constituted by 'new' goods and if US imports were mainly constituted by 'mature' goods.

Given the importance the Leontieff paradox had in determining the emergence of new technology theories of international trade it is quite interesting that Leamer (1980) has recently challenged the validity of the procedure used by Leontieff to demonstrate the existence of the paradox. According to Leamer under different and more correct assumptions the paradox is reversed. Subsequently however Stern and Maskus (1981) using Leamer's criterion demonstrated that the Leontieff paradox did hold for some years and was reversed for others. These uncertainties about the validity of the Leontieff paradox do not detract from the significance and degree of empirical validation of new technology theories of international trade (Soete, 1981).

Another related aspect of international trade which has been fairly well documented empirically and which is not easily compatible with comparative advantage theories is the importance of non-price factors in international trade. Comparative advantage meant an advantage which allowed a country to produce a good more cheaply. However, a considerable amount of evidence has been gathered recently that sometimes more expensive products outsell cheaper ones, therefore pointing to the importance of non-price factors in international trade (Posner and Steer, 1979; Stout, 1977). For example, unit values of West German and French exports tend to be higher than UK exports in a wide range of goods. Evidence also exists concerning the importance of packaging, distribution, advertising, delivery times and so on in influencing competition. On the whole, then, the factors influencing competition can be divided into three types: price, *intrinsic* non-price factors (those which are *embodied* in the product) and *associative* non-price factors (those factors which depend on the organisational arrangements adopted in the production and distribution of the goods) (Saviotti *et al.*, 1980; Gibbons *et al.*, 1982).

Clearly technology plays a very important role in conferring unique characteristics to a product and this is especially true in the case of a multicharacteristics product (Chapter 5). To the extent that non-price factors can be related to differences in technology they fall naturally within new technology theories of international trade. However, as shown by the distinction between intrinsic and associative non-price factors, the associative non-price factors cannot be equated with technology, at least not within a narrow definition of it. Such factors will depend on the total knowledge base of firms and on its relationship to firm organisation. As will be shown later, this knowledge base is closely related to the intangible assets which provide an important explanation for the existence of multinational enterprises. Further discussion of this point is therefore deferred until the section on multinational enterprises.

The Leontieff paradox and the related importance of non-price factors are by no means the only problems faced by comparative advantage theories of international trade. Some assumptions of the theory, such as perfect competition, constant returns to scale, factor immobility, the immediate and free diffusion of technology, are known to be very unrealistic. Furthermore some welfare implications of trade, such as the worsening of the terms of trade and the disappearance of some national industries are at odds with comparative advantage theories. In practice innovation and technical change are themselves elements of imperfect competition and the tendency to reestablish imperfect competition in the presence of the international diffusion of technology is one of the most powerful factors influencing foreign trade.

The importance of technological innovation and change in influencing foreign trade is by now acknowledged. A debate exists about whether technological change can be accommodated within modified comparative advantage theories or whether more radical departures, such as those represented by new technology theories, are required. The comparison is somewhat uneven because, in spite of their greater intuitive appeal and empirical validation (Soete 1981), new technology theories have not achieved an analytical status comparable to that of the more traditional theories.

It seems likely that this debate will not be resolved without some more explicit treatment of the interactions between intra-firm trade and inter-country trade. We now turn to the issue of trade and MNEs.

The growing importance of multinational enterprises can be described as a change in the mechanisms by which trade flows occur. Economists and organisation theorists trying to explain the rapid growth of MNEs had to explain why setting up wholly or jointly owned subsidiaries in foreign countries to manufacture particular products was for many firms preferable to exporting from their home base country. In principle there are reasons both for and against this type of behaviour. For example scale economies and the greater familiarity of foreign firms with their local market would put MNEs at a disadvantage with respect to their local competitors and therefore discourage foreign direct investment (FDI). On the other hand cheaper labour and high import taxes in foreign countries could provide positive inducements for firms to invest in manufacturing facilities abroad. The fact that intra-firm trade has grown faster than inter-firm trade seems to imply that under most circumstances the advantages of foreign direct investment outweigh those of arms length transactions. An aim of a theory of FDI is therefore to compare systematically its advantages and disadvantages.

A considerable literature exists about MNEs which cannot be reviewed fully here. The reader is referred to the work of Caves (1982) for a review. Our discussion is restricted to some aspects of MNEs' behaviour which are influenced by and influence technological change.

According to Caves (1982) one of the most fruitful concepts in explaining the existence of MNEs has been that of intangible assets. An intangible asset could be something like technological or marketing knowledge, which allows a firm to produce a product that none of its competitors can exactly match or which gives a firm a unique knowledge of consumer demand. The concept of intangible assets provides an explanation of why a multiplant firm with central ownership and co-ordination is in some way more efficient or profitable than the same plants managed by independent firms. An intangible asset yields the firm possessing it a rent, but this rent cannot be realised by means of market transactions given the nature of intangible assets.

Some examples of intangibles are public goods. Once they have been produced the marginal cost of their use is zero or close to zero and so they should be sold at a zero price. If this were the situation intangibles would simply not be produced. Some institutional way has to be found to reward the producer of public goods. An example of such an institutional approach located outside the firm is the patent system, which gives the inventor of a particular technology a temporary monopoly over the technology itself. Any user of the technology has to pay royalties and fees to the inventor until the patent expires. If the transactions involving the use of an intangible asset occur within the same enterprise there is no conflict between production and use of the asset itself.

Caves follows Williamson (1981) in arguing that intangible assets suffer from *information, impactedness, opportunism* and *uncertainty*. A firm which has developed a new technology has to disclose its nature in order to be able to sell it, but if the nature of the new technology is disclosed too completely then the buyer might simply use that information without having to pay for it. In other words, there is a minimum information content required to be able to use a technology (impactedness) and both buyer and seller suspect the other party of opportunism. Furthermore, neither buyer nor seller know exactly how useful the technology is going to be (uncertainty) and it is therefore difficult to attach a correct price to it.

Given these characteristics of intangible assets, arms length transactions are likely to be more complicated and less efficient than intra-firm transactions. Obviously this line of explanation is nothing other than an extension of transaction cost analysis (Chapter 2) to the international arena. Clearly within this type of explanation the existence of MNEs is closely related to the technology they use. However, it is not only technological knowledge in the strict sense of the term which can constitute intangible assets. Managerial, organisational and marketing knowledge can be equally important. The nature of these intangible assets is such that once a firm has acquired them the 'capacity' to use them is not limited. In other words, a firm can accumulate an excess managerial capacity (Penrose, 1980, Ch. 2), which can constitute an incentive for further expansion both nationally and internationally. It is particularly important to realise that a firm's managerial or technological capacity is not the collection of a series of individual skills but the resource of the organisation or 'team'.

Intangible assets are closely related to enterprise knowledge and they make a very important contribution to the explanation of the existence of MNEs. Technological knowledge is an important source of intangible assets although not the only one. One could therefore expect MNEs to be important producers of technological knowledge and to be concentrated in knowledge intensive industries. That MNEs are important producers of technological knowledge is borne out by empirical research. Firstly, a close correlation has been found between R & D spending and the share of foreign subsidiaries in the assets of US corporations (Caves, 1982). Secondly, Mansfield, Romeo and Wagner (1979) found that US companies expected to obtain a significant proportion of the returns from their R & D projects from overseas markets. R & D is therefore not only one of the causes of companies' internationalisation but it is itself affected by existing patterns of international expansion of MNEs.

The evidence that MNEs are preferential users of R & D shows that they make intensive use of one of the inputs required to generate new technologies. There is also evidence that they are associated with the production of new technologies. The product cycle does not explicitly try to explain the existence of MNEs but only to explain the types and directions of world trade flows. However, on the basis of it one would expect that since the US is the most innovative country in the world and the largest R & D spender it should be the home of most MNEs. This is indeed borne out by experience and reinforces the connection between generation of new technologies and enterprise internationalisation.

Perhaps even more important than their role as producers of new technology is the contribution of MNEs to technology transfer. The many possible ways in which technological knowledge can be transferred fall between the two extremes represented by licensing to an independent firm, an example of arms length transaction, and intra-firm transfer. Both these modes of transfer have their advantages and disadvantages. For example licensing to an independent firm will be convenient when the licensor lacks some assets other than the intangibles which are required for FDI, such as capital, an extensive sales and assistance network and so on. Also, licensing decreases the risks deriving from changes in the foreign country where investment takes place (for example expropriation) and it has a shorter lead time than starting a subsidiary from scratch. On the other hand intra-firm transfer avoids any leakage of technology to other firms and it is more convenient than licensing when arms length transactions are complex and difficult to enforce. Another situation in which intra-firm transfer is convenient with respect to licensing arises as a consequence of the transfer process itself. That transfer costs were negligible with respect to production costs was until recently held to be an important imperfection in the market for knowledge. However, Teece (1977) showed that transfer costs can be considerable and can in particular situations account for more than 20 per cent of the cost of developing the technology. Moreover, transfer is easier and therefore less costly when knowledge is *codified*, or expressed in an abstract and symbolic form, than when it is *tacit*. Intangible assets, or at least the most strategically valuable parts of them, are likely to be tacit and according to Teece (1981) the costs of transferring them are likely to be lower for intra-firm transfer.

In a related way it has been argued (Magee, 1977) that MNEs transfer technologies which are more *appropriable*. Appropriability could either be an intrinsic property of the technologies transferred or of the way in which they have been transferred and it is probably due to both. Tacit knowledge is clearly more appropriable and intra-firm transfers reduce risks of technological leakages.

The existing evidence that intra-firm trade has been rising more rapidly than inter-firm trade seems to imply that it is altogether advantageous with respect to licensing. Of course this conclusion might be applicable only to some industries or technologies which would be those characterised by a high incidence of MNEs. If this interpretation is correct it implies that it is more 'convenient' for a firm to set up a subsidiary in a foreign country rather than to license the technology to a potential competitor in the same country. However, in any instance of technology transfer two parties are involved, the supplier and the recipient, and the previous considerations only prove that intra-firm transfer is more convenient for the supplier. There is considerable evidence that recipient countries are not as enthusiastic about the advantages of intra-firm transfers and that they have tried to limit what they perceive as negative consequences of FDI (Lall, 1980; Bastos Tigre, 1982). This is a very important research area, but one which belongs to the specialty of economic development and which therefore cannot be adequately reviewed here.

Summarising what has been said so far, new technology theories of international trade tend to emphasise technology as the crucial variable which determines commercial flows as opposed to comparative advantage theories which stressed factor endowment. But both of these theories are concerned only with commercial flows and not with the institutional context in which trade in goods and technological knowledge takes place. On the other hand theoretical and empirical analysis of MNEs has provided valuable insights into the reasons for which intra-firm transactions have become more frequent than arms length transactions. These two lines of explanation belong to different research traditions and a complete synthesis of the research on foreign trade and of that on industrial organisation has not yet been achieved.

An interesting attempt in this direction has been made by Dosi (1982, 1984). In his view the tendency to the formation of oligopolies is natural, at least in many industrial sectors, and technology plays a fundamental role in determining the path towards oligopoly. A producer who has developed a new technology acquires an initial advantage which is subsequently amplified by the cumulative character of technological knowledge. Certain technologies, for example those characterised by a high rate of technological change, high appropriability, static and dynamic scale economies, and so on, lend themselves better to explain this pattern of growth. The technological leaders which emerge in this way form a national oligopoly. This stage is followed by the unification of world markets which subsequently leads to the formation of an international oligopoly. During this process of expansion at a national and international level there are important processes of interaction and feedback between technology and industrial organisation but technology can become, at least during certain stages of this evolution, the most important factor. This is a further example of the reverse causality already encountered in analysing the relationship between innovation and market structure (Chapter 5). It was then found to be necessary to introduce the concept of technological opportunity to explain the different patterns of invention and innovation of different industrial sectors. An extension of that argument now leads us to explain patterns of international industrial expansion on the basis of an enlarged concept of technological opportunity. It therefore follows that this is a further example of the need for technological change to be analytically separated and to see aggregation by technology as an alternative approach to aggregation by economic units.

The preceding arguments may explain why some firms, having accumulated specific advantages, can expand faster than other firms, but it does not explain the location of their international expansion. Countries which become hosts to MNEs must have some country specific advantages (Dunning, 1979, 1980) which attract FDI in preference to other locations. Such an 'eclectic' approach combines the analysis of industrial organisation and that of commercial flows and comparative advantage. It may therefore be able to bridge the gap between these research traditions.

6.9 CONCLUSION

This chapter has revealed that the relationships between technical change and the macroeconomic domain of output, employment and trade have been analysed within a bewildering array of research frameworks. Following the initial emphasis of classical economists on the enabling role of technological change, neoclassical analysis tended to abstract from technical change. The early attempts to reintroduce it used derived indicators rather than explicit measurement or conceptualisation. Methodological borrowings from behavioural theory of the firm, and studies of industrial organisation, have reintroduced a richer concept of technological change, but often at the expense of formal analysis. Sectoral interactions, with explicitly stated differences in technology and demand, have revealed the danger of additive aggregation. It would appear that future progress in the field must be achieved in a context in which formal and narrative modes of analysis can be reconciled, and in which analyses at the level of firm, industry and economy are made progressively more mutually consistent. The continued growing importance of MNEs must also alert us to the need to recognise the validity of non-economic disciplines in giving richer accounts of the 'intangible assets' which influence corporate behaviour and resulting economic aggregates.

7 Technology and Structural Change: Output and Employment in the Long Run

7.1 STRUCTURAL CHANGE

Earlier chapters have followed the conventional procedure in economic analysis of distinguishing between microeconomic and macroeconomic problems. Thus in Part I the focus was on the firm as the initiator of technological change. Technological changes brought about by other firms were related to those of the firm under consideration by translating them into components of the environment, which might further be broken down into threats, opportunities, changes in the structure of demand, in patterns of risk facing various lines of technological advance, and so on. Chapter 5 examined the evidence for patterns amongst innovations. By contrast, Chapter 6 dealt largely with technological change from the 'top down', rather than from the 'bottom up'. By assuming that significant numbers of firms are always innovating or adopting, it is possible to hypothesise an average 'rate of technical change'. On this basis it has been possible to construct economic arguments about the likely effects of a rate of technological change on rates of productivity growth, output growth, trade patterns and so on.

These lines of argument have produced useful insights, but as has been argued above, they are also beset by problems. A particular difficulty is that of making the assumptions of the micro and macro approaches consistent with each other. The most obvious example of this is the aggregation assumptions involved in much macro level analysis, such as that on rates of economic growth discussed in the preceding chapter. It is no less true of micro level analysis however. The discussions of firms' activities in technological innovation often assume that the macroeconomic environment is unaffected by individual firms. This is clearly not always true, especially in the case of large MNEs.

There is therefore a pressing need for methods of linking together the micro and macro level analyses of technical change and of economic phenomena in general. There are, in our preceding discussions, some pointers as to where such a linkage may be found. Three of them are worth mentioning here. Firstly, it has been shown that innovations, once on the market, tend to follow a reasonably well ordered diffusion path, in which relevant constituencies of adopters and suppliers can, to some extent, be identified in advance. The diffusion process has been shown to be susceptible to microeconomic analysis (see section 5.6) of a quite explicit nature. It is also clear however, that if the innovation is of sufficient size, the macroeconomic aspects of diffusion processes may come to the fore in the form of incomes to capital and labour, employment effects, forward and backward linkages, substitution effects, comparative trade advantage and so on. Secondly, and closely related to diffusion, the concept of the product life cycle and its various refinements provides further connections between micro and macro levels of analysis. Preceding discussions of this topic have demonstrated its attempt to encompass factors as diverse as patterns of international trade (Vernon, 1966) and firms' strategies and organisational characteristics (Abernathy and Utterback, 1975). Thirdly, the discussion of science-push and demand-pull influences on innovation is relevant here. The discussion in Chapter 5, and in particular the work of Walsh *et al.* (1984) on the chemical industry and its sub sectors has shown that, if one chooses the appropriate level of aggregation for analysis, certain historical patterns in the operation of these two influences may be found. The early stages of the development of some of the chemical industry's major branches were shown to be more often displaying science-push phenomena, whereas a more intricate balance between science-push and demand-pull characterised subsequent developments. The historical analysis of *the sectors* was a pre-condition for resolution of the more general analytical problem of relationships between science and demand.

These three potential contributions to a link between micro and macro level analysis: diffusion processes, product cycles, and sector level patterns in the interaction of science-push and de-mand-pull factors, share an important common feature. This is their ability to utilise both the industry or market level of aggregation, which is the traditional domain of industrial economics, and the technological level of aggregation which we have stressed at several points in this book. We believe that this approach is a fruitful one, and it lies behind the remainder of this chapter.

In addition, the analysis of Pasinetti summarised in the preceding chapter, strongly suggests that differing rates of technical change and of growth of demand across industries will be the principal forces generating aggregate output and employment levels, and that equilibrium or balanced growth will be unlikely (see in particular Pasinetti, 1981 Ch. 10). This gives further support to the importance of the industry level of analysis as the nexus of the micro and macro perspectives, rather than simply a topic in its own right.

These arguments throw a new light on one of the issues which has long been a preoccupation of economists working at the industry level, namely the 'productivity puzzle', or persistent dispersion of productivity growth rates across industrial sectors.

Over any reasonable period of time (say a decade or so) it becomes clear from computations based on Census of Production data that there is a wide dispersion across industries of rates of growth of output, rates of growth of productivity, rates of growth of employment, and rates of price change. Some industries have productivity or output growth rates two or three times greater than the average, while others will only have rates half that of the average. This suggests that quite different conditions prevail in these industries, and it has been an important problem for economists to explain what these are and how they affect productivity and output. This is the phenomenon which is sometimes known as the 'productivity puzzle'. Its relevance to structural change is quite clear. If, over a considerable period, some industries grow very much faster than others, then by the end of that period, their proportional share of output, capital, employment, profit and so on, in the total economy will be much greater than it was at the beginning of the period. It is this process which makes an economy change in shape and character at the same time as growing in quantitative terms. It is clearly of the utmost importance in all manner of ways which *particular* industries grow most quickly and which grow slowly or decline. Many problems of analysis and policy are ultimately related to this issue.

In the case of the UK economy we are fortunate in having a number of studies on differential productivity and output growth which cover most of the last sixty years. The landmark study of Salter (1966) covered the period 1924 to 1950, and the post-war period has been analysed by Wragg and Robertson (1978) and by Freeman *et al.* (1982, Ch. 7). These two periods differ in some quite important respects because of the quite different rates of overall growth which they display. This point is discussed later in this chapter in the section on long waves. However, in a number of ways the data of the two periods support some quite important general arguments concerning the relationships between technical change, productivity change, price change and output change at the industry level. These arguments deserve review.

Let us take first of all the primary phenomenon of variation in industry rates of productivity growth. Salter (1966) examines three possible causes; these are firstly variations in the efficiency of labour, secondly variations in the rate of substitution of capital for labour, and thirdly variations in the rate of technical change. The first two of these are undermined both by data and by argument. The variations in earnings and labour costs do not match those in productivity growth rates, and it is intuitively unlikely that labour efficiency could vary by the amount necessary to generate the large variations in labour productivity growth. Similarly, Salter shows that the variations in rates of factor substitution do not correlate with rates of productivity growth, though he does not rule out the likelihood of substantial substitution accompanying technical change (Salter, p. 132). It is this last variable, technical change, which Salter therefore identifies as being the principal explanation of variations in productivity growth. This is a view shared by Wragg and Robertson, and Freeman *et al.*

This is an important hypothesis. First of all it is a view of industrial growth which is very supply-side biased by comparison with conventional economics, and this point is underlined when we consider the link between productivity, prices and output below. Secondly, the view that the 'rate of technical change' is intrinsically very variable across industries contrasts with the traditional macroeconomic assumption that exogenous technical change is 'on average' continuous and evenly distributed. If indeed the rate of technical change varies across industries, it is worth enquiring what the source of this variation might be. Salter himself advances the view that some industries are more 'scientifically based' than others, and therefore possess greater capacity to benefit from technical change. He adds that the 'age' of an industry is a contributory factor in that the central technical features of the processes may be more or less susceptible to improvement. Since Salter's work, other authors have offered more elaborate variations on these themes. In Part I of this book we have discussed Nelson and Winter's (1977) notion of natural trajectories of technological change; Kay's (1979) notion of 'technological opportunity' as a significant variable explaining research intensity at industry level; elaborations of this idea based on contributions from several other authors; and Chandler's (1962) analysis of the development of large corporations and vertical integration which placed emphasis on the 'growth potential' of certain industries as a result of their basic technologies and the scope for improvement and integration. All of these developments can be seen as more developed forms of Salter's simpler view, which support the basic thrust of the argument.

It is also important to acknowledge the role of demand however, and its interaction with technical change. Chandler's analysis not only emphasised technology, but also the rate of growth of the markets facing the rapidly growing industries. Similarly, other economists who wish to emphasise the role of demand have pointed out that the variations in technical change and productivity may in fact be caused by differing rates of demand growth facing different industries. (Kaldor, 1966).

In fact, Kaldor's model introduces another of the basic relationships examined by Salter, namely the relationship between output growth rates and productivity growth rates. Kaldor argues that demand growth leads to output growth, which leads to economies of scale and therefore to productivity growth. Salter's view is that productivity growth (which is 'technology-pushed') leads to falls in the relative price of output, which then leads to demand and output growth. What this difference amounts to is two alternative causalities in the explanation of Verdoorn's law (the observed correlation between industry rates of output growth and rates of productivity growth). One causality runs from technology, supply and productivity, via prices, to output growth (Salter); the other runs in the opposite direction (Kaldor). In many ways these views mirror at the industry level the more general controversy already referred to (Chapter 5) concerning the relative roles of science-push and demand-pull in innovation. We have already argued that in the latter case the situation is characterised by interaction between these two forces with the former perhaps dominating the latter in the early stages of *the growth* of an industry or sector. Therefore it would be consistent to make a similar judgement in this case where we are analysing not just innovation *per se*, but its effect on productivity, prices and output.

We can therefore conclude that both Salter's and Kaldor's explanations capture different features of reality, and that the Salter mechanism is most appropriate in explaining growth in those industries, and at those times, where technical change, and particularly radical technical change is important. However, in the case of some industries, (for example, textiles or clothing) it is clear that total demand does not increase whatever the effect of technology on relative prices, and here a simple supply-based model breaks down. Hence we must further conclude that statistical cross-section tests of alternative relationships between technological change, productivity change, and output growth must be tempered by specific analysis of the real historical features of the industries, the nature and purpose of their outputs, the effects of product innovation as well as process innovation, and the interactions between industries. In this way the full context of structural change might be discerned; it cannot be read off directly from varying rates of productivity growth.

It is useful to recall at this point the arguments of Chapter 5 on technical change and market structure. So far we have discussed the possible links between technical change and structural change across sectors, but it is also important to register the effects of technical change on structural change within sectors. We argued in section 5.4 that, while the evidence on market structures in *general* is ambiguous with respect to its effect on innovation, and with respect to the inverse effect, it does seem clear that within one sector increasing concentration is related to progressive technological maturity. However, set against this are the phenomena of diversification by large firms, small firm entry with radical innovation, and technical convergence, all of which tend to continuously blur the edges of 'sectors' and re-emphasise the importance of the former type of structural change. Nevertheless, increasing concentration in 'mature' industries seems to be a clear trend in the twentieth century. (See also section 7.3 of Freeman *et al.*, 1982, for a discussion of the Schumpeterian position on tendencies of increasing concentration).

The fact that the definition of an industry and changes in its boundaries is important both for analysing actions of business units within that industry, and also for analysing interactions with other industries, further underlines the basic point that the sector, and structural change defined in terms of sectors, plays a vital integrating role in linking together micro and macro-levels of analysis. In the remainder of this chapter, the idea of structural change in used to analyse changes in output and employment in the long run. A discussion of economic growth based on the view that technical change is *not* uniform across sectors, and that it acts via structural change rather than directly, is based on quite different premises from the residual or production function approach outlined in Chapter 5. The underlying approach is one of trying to identify particular phases or periods of change in which discernible *patterns* in the interactions of industries, markets, factor supplies, and institutional climates create characteristic patterns of growth. The method is therefore a combination of analytical and historical and empirical work.

Many authors have given accounts of the growth of capitalist economies framed in these terms, and they have certainly not all come up with the same results. Cornwall (1977, Ch. 3) gives a useful comparison of this style of analysis with conventional growth theory, and the remainder of that volume presents an account of growth in which the flexibility of economies defined in terms of their supply of inputs is the crucial variable which 'permits' utilisation of available technology and high levels of structural change and growth. In contrast, Maddison (1982) offers an account in which major 'shocks' to the economic system at random intervals give an impetus to an economy to depart in one direction or another from notional neoclassical equilibrium growth paths. While both of these accounts have much to commend them, our view is based on a third style of analysis, which makes use of a re-developed version of the 'Kondratiev long wave' theory, suggesting that the world capitalist economy has exhibited a series of alternating phases of accelerated and decelerated growth, each lasting twenty to thirty years. It is argued that these phases are related to technical changes and structural changes which have certain predictable dynamic behaviour patterns, and institutional changes which do not. These arguments are reviewed at length in the remainder of this chapter.

7.2 LONG WAVES OF ECONOMIC GROWTH

Ever since long waves were first proposed as a way of interpreting the uneven growth of the world economy in the early part of the century there has been controversy not only over the explanatory mechanisms put forward but also over the data needed to show that such waves exist. Kondratiev, the Russian economist whose name has become associated with long waves, spent much of the early period of his work on the topic in the 1920s simply assembling the long runs of data on prices, production, investment, interest rates and employment which he needed to argue that the waves were a real phenomenon. (Kondratiev, 1978). The construction of these data series is beset with many problems and, therefore, many writers have little confidence in what can be reliably deduced from them. Furthermore, even today, sixty years after Kondratiev, we have only experienced four complete long waves in the industrial epoch, which makes identification and explanation in anything approaching a cyclical manner a very risky business. Nevertheless, as the slowdown in economic growth in the west has continued during the seventies and eighties, more and more evidence has been assembled which suggests that it may be a long wave downswing similar to those of the 1930s and 1880s. Before proceeding further with explanation and interpretation, it is appropriate to summarise some of the data.

Gordon, Reich and Edwards (1982) in their analysis of long waves in the US economy, have usefully collected together a variety of indicators which demonstrate the basic proposition of alternating periods of faster and slower growth. Some of that data is reproduced below.

Table 7.1 shows average annual percentage growth rates of real output for four countries. In every case except for that of Germany in the inter-war period the direction of change is that which

TABLE 7.1   *Growth of real output over the long swing*

Long swing	Years	Average annual percentage growth in real output				
		United States	United Kingdom	Germany	France	Weighted average
IIA	1846–1878	4.2	2.2	2.5	1.3	2.8
B	1878–1894	3.7	1.7	2.3	0.9	2.6
IIIA	1894–1914	3.8	2.1	2.5	1.5	3.0
B	1914–1938	2.1	1.1	2.9	1.0	2.0
IVA	1938–1970	4.0	2.4	3.8	3.7	3.8

Note: The Roman numerals refer to the number of the long wave, according to Gordon, Reich and Edwards' chronology. The letters A and B refer to the 'A-phase', or accelerated growth period, and the 'B-phase', or period of slower growth.

Source: Gordon, Reich and Edwards (1982) *Segmented Work, Divided Workers.*

TABLE 7.2   *Expansion and contraction over the long swing*

Long swing	Years	Expansion/contraction ratio (months)		
		United States	United Kingdom	Germany
IIA	1848–1873	1.80	2.71	1.61
B	1873–1895	0.86	0.76	0.79
IIIA	1895–1913	1.14	1.62	1.33
B	1919–1940	0.67	1.36	1.82
IVA	1948–1971	1.95	n.a.	n.a.

Source: Gordon, Reich and Edwards (1982) *Segmented Work, Divided Workers.*

supports the long wave interpretation. As Gordon, Reich and Edwards point out, the inflationary boom of the Weimar period, followed by the special economic circumstances of Nazi expansionism may explain Germany's departure from the pattern during this period.

Table 7.2 deals with the interaction between the long wave and the more familiar short term business cycle. These latter cycles are

TABLE 7.3   *Growth in world trade over the long swing*

Long swing	Years	Average annual percentage change in world trade
II A	1850 to 1876–80	+ 7.04
B	1876–80 to 1891–5	3.75
III A	1891–5 to 1913	5.42
B	1913 to 1936–8	0.26
IV A	1936–8 to 1970	10.98

Source: Gordon, Reich and Edwards (1982) *Segmented Work, Divided Workers.*

superimposed on the long waves, but they tend to be skewed in a way which reflects the underlying expansionary or contractionary impetus of the long wave. Thus in long wave upswings, business cycles tend to have strong booms and short, weaker than average depressions. Conversely, in the long wave downswings, the business cycle depressions tend to be more severe, and the booms tend to be weaker and more short-lived. The data show the ratio of expansionary to contracting months over the long swing, which is an indicator of the process just described. Again the evidence supports the possibility of a long wave pattern, with the exception of the case of Germany in the twenties and thirties, already mentioned.

Finally, in Table 7.3, evidence is presented on the growth in the volume of world trade over the long wave. Since most authors see the long wave as being fundamentally a feature of the world economy, in which different national economies participate to different extents as a result of all manner of contingent factors, then this international indicator is of particular interest. Here the evidence seems very suggestive indeed, with changes of almost an order of magnitude in the annual percentage growth rates between some long wave phases.

Despite its suggestive character the limitations of this data mean that it would be impossible to 'prove' in a strict sense the existence of any rigid cyclical mechanism. For this reason arguments about whether a mechanism of such a rigid nature exists or not are somewhat beside the point. We can agree with Freeman (1984) however, in taking the view that the 1880s, the 1930s, and the

1980s have been periods of severe structural crisis in the world market economies. During these periods there has been the paradoxical situation of some industries in decline or decay, alongside others showing dynamic growth potential, together with high levels of unemployment, and often significant tensions in the social and political institutions most closely related to the functioning of the economy. These crises, which raise the issue of structural change both in the specific sense used earlier in this chapter, and in the broader sense of the social and institutional context of industrial structure, have been of varying length rather than of any fixed period. But none has been less than two decades and all have spanned more than one conventional business cycle. Between the periods of crisis have come long but varying periods of relative prosperity and economic growth. For this reason we would argue that there is an important phenomenon to explain. It clearly demands a very multi-faceted style of explanation, and will not sit easily within the normal approach of economic analysis, which tends to abstract from, or assume constancy of, the very structural and institutional parameters whose long-term evolution may well be the most important determinant of the economic phenomena. The authors whose work has contributed to this project have utilised historical, technological, sociological, political and economic styles of analysis in various ways. While the result may well be eclectic it is nonetheless revealing.

7.3 NEO-SCHUMPETERIAN THEORIES OF LONG WAVES

Since our focus in this book is on technical change, it is appropriate to concentrate on those theories of long waves which have placed technical change most centrally in their accounts. These are the theories of Mensch (1975) and Freeman *et al.* (1982). The basic Schumpeterian approach which they develop can be recapitulated briefly from the now well-known 1939 volume. Schumpeter believed that one of the most fundamental features of capitalism was its tendency to disequilibrium. This resulted from the ever-present possibility of entrepreneurs' seizing upon inventions, which in his early work he largely regarded as exogenous, and turning them into innovations. Such an innovation would, in his view, carry with it temporary monopoly profits which would attract a retinue of imitators and begin what we would recognise in today's terms as a diffusion process. However, as has since been formalised by Metcalfe (see section 5.6) the monopoly profits are gradually competed away and the market for the innovation eventually reaches some equilibrium value via a process of retarded growth. The incomes, derived demands, and stimuli to other economic activity created by this diffusion process can, in principle, create a net increase in aggregate economic growth in the system as a whole which will also have the characteristic sigmoidal shape of the diffusion curve (this assumes a simple model of one innovation and equilibrium in all other branches). At the end of the diffusion process, in Schumpeter's view, entrepreneurs' confidence and propensity to invest, both in the new branch and in general, may be depressed in an exaggerated way because where and whether new innovations will emerge are uncertain. The net effect on the economy can thus become deflationary. What is proposed here is the familiar cyclical pattern of a 'virtuous circle' of investment and demand growth being replaced by a vicious circle of declining investment and declining demand growth. The trigger factor in this case is the effect of new technical possibilities upon business expectations and therefore upon the 'animal spirits' component (Keynes, 1936) of aggregate investment.

Schumpeter tried to base his explanations of cycles of all lengths on a mechanism of this type; explaining the differing lengths by recourse to the existence of innovations with differing degrees of importance, scale and diffusion time. This is really the crux of Schumpeter's argument (Schumpeter, 1964). There is no reason to question the first part of the account; clearly the introduction of a new product and its evolution into a mature branch of production can create new increments of income to capital and labour, and therefore give a once-for-all increase to economic growth. But for this process to be *cyclical* the innovations have to spaced out in time, and either clustered together in groups or else of intrinsically large size. If they are not structured in this way but merely randomly spread in size and time, then the net effect on aggregate output would be an indeterminate pattern of fluctuations with no cyclical character. Rosenberg and Frischtak (1983) follow several of Schumpeter's critics in questioning whether innovation clusters and their causes can be demonstrated.

Schumpeter certainly believed that clusters of innovations did occur. In particular he felt that the very process of diffusion of an initial innovation, if it was of sufficient importance, would generate further related innovations because the 'retinue of imitators' would tend to improve upon the first innovation and to create other innovations in related products, processes, technologies and organisational structures (Schumpeter, 1964). This view of exactly what is meant by a 'cluster' of innovations is critical to the later stages of this argument. We shall see that Rosenberg (1976, 1983) and Freeman (1984) have developed it to a greater degree of clarity. First however, we shall examine briefly an important earlier attempt to test Schumpeter's clustering hypothesis which has proved very influential in stimulating the long wave debate since the mid-70s.

Mensch (1975) presents lists of innovations compiled by other researchers covering various periods of the twentieth and late nineteenth century. Neither he nor the originators of the lists claimed any representative status for them; it is anyway difficult to imagine how such status could be demonstrated. He does argue, however, that they are 'basic' innovations which are in some sense more important than other innovations. Also, because these lists had been collected, studied and *dated* by other researchers for other purposes he argues that they are a sound sample on which to conduct a test for clustering. His results suggest that the innovations are indeed clustered in the deepest part of the long wave depressions rather than randomly distributed in time, and he takes this as strong evidence for a long wave phenomenon in which innovations play an important, even a determining role.

Mensch proposes the following explanation for the clustering process. As the impetus from one set of innovations and new branches of production begins to subside and the upswing reaches its peak, firms begin to engage in 'pseudo' innovations (product differentiation) which they see as possibly protecting their share of the market though not expanding the market as such. But this behaviour tends to crowd out potential basic innovations, even though it is basic innovations which are necessary to get the system moving again. In other words, he suggests a conflict between the micro and macro economic rationalities. The result is a deepening of the depression, a sharpening of competition, and a further worsening of the prospects for any radical or basic innovation.

This is another dimension of the familiar vicious circle metaphor mentioned earlier.

The critical step in Mensch's argument comes next when he argues that sooner or later the depression exerts such severe pressure upon some entrepreneurs that the balance between risk and incentive shifts back toward incentive, and they introduce basic innovations because they see innovation as the only means of survival when existing markets are stagnant or declining. (It should be added that it would be entirely consistent with Schumpeter's views to add to this mechanism the possibility of new small firms emerging to pioneer the basic innovations provided that they can muster the resources necessary.) This is the 'depression-trigger' effect on basic innovations which is the core of Mensch's argument. It should be noted that he is not arguing that there is any alteration in the rate of inventions; these he sees as being much more exogenously determined. But rather there is in his view a foreshortening of the lead time between invention and innovation brought about by the depression. The net effect is a new cluster of innovations which open up new markets, new opportunities for profitable investment, and which form a core for the development of a new long wave upswing. In summary then, Mensch sees the long wave as caused by the synchronising of a number of new-product life cycles which are large enough in aggregate to stimulate the whole economy. The downswing is characterised as a 'technological stalemate' which is eventually resolved by a further burst of innovations.

The other principal neo-Schumpeterian theory of long waves is that of Freeman, Clark and Soete (1982). Whilst they also see life-cycles of new sectors as a powerful force in the long wave, they also introduce an important role for the labour market, and they have a quite different approach to the question of the role of technical change in the long wave. They disagree strongly with Mensch's depression trigger described above. Firstly, they question the data which Mensch has used to search for clustering of innovations on the grounds that the samples themselves are too *ad hoc* to present any adequate coverage of the major industrial fields involved in the long wave booms; the dates of the innovations given by Mensch are open to disagreement, yet they are obviously critical to the presence or absence of clusters; and there is no adequate definition of 'basic innovation' to ensure that those innovations in the sample are a sensible sub-set of innovations in general. While this does not disprove the clustering hypothesis, it does in the view of Freeman *et al.* make it impossible to support it with analysis of the data presented by Mensch (Freeman, Clark and Soete, 1982).

Secondly, they question the plausibility of the depression trigger hypothesis used by Mensch to explain clustering. Since some of Mensch's data are innovations which were the subject of case-studies (Jewkes, Sawers and Stillerman, 1958) it is possible to check to some extent whether, as Mensch's theory would require, the depression played any motivational role in the innovations, and whether the invention–innovation lead times were in fact shortened. Freeman *et al.* find no evidence for either of these hypotheses in the case study material. Instead they argue that the case-study material suggests that the only factor to emerge as common stimulant in some of the 1930s innovations is the expectation of increased demand associated with re-armament. This is not to argue that individual recession-induced innovations are impossible; clearly they are possible. Rather it is to suggest that they will not be sufficient in number, or of the appropriate character, to fuel a new long wave boom. In general the view of Freeman *et al.* is that R & D activity is decreased in long wave depressions not increased as Mensch's theory might suggest. (pp. 51–57). Furthermore their own analysis of some data on patents over a 200-year period suggests that clusters exist during long wave booms as well as recessions. (Patents are normally taken to refer to inventions rather than innovations, though in fact they have a more variable relation to date of innovation. In this case Freeman *et al.* use the patent data as only an indirect proxy for examination of possible clustering of innovations).

These criticisms of Mensch's treatment of the 'lower turning point' of the long wave are convincing. If we are to retain the notion that new products and sectors create long wave booms through a Schumpeterian process of structural change then a more satisfactory account is needed of how and why the particular new products and sectors emerge at the appropriate time, and why their eventual approach to maturity is not immediately followed by a new set of products. Freeman *et al.* offer an alternative account which is centred on the concept of 'New Technology Systems'. There are two essential features of the New Technology System idea (NTS) which differentiate it from a cluster of basic innovations. Firstly it consists of technologies which are very widely applicable in many products and processes in many industries, thus generating a range of related innovations. Secondly it is argued that it is the clustering of diffusion processes for these innovations rather than clustering of the dates of the innovations themselves which is both the appropriate interpretation of Schumpeter's position and the most important stimulus to the long wave upswing. These points deserve further discussion.

If technology is to play a part in a theory of long term growth, then it does seem reasonable to argue that something as substantial and historically specific as a long wave upswing will have distinctive and quite specific technological characteristics rather than being the result of an essentially random and unrelated collection of innovations. Structural change, if it is to be made a principal cause of economic change ought to be describable in terms of a direction as well as a magnitude. This implies that it is necessary to examine the specific technologies which have, during any period of sustained economic growth, shown the most capacity for generalisation and penetration of a wide number of different products and processes. Freeman *et al.* argue that these technologies act as threads which tie together the progress of different industries and firms through the media of specialist suppliers of materials and components and special skills. Thus there may be a cumulative process of technology transfer across sectors in which each improvement in the use of a new set of technological practices may generate benefits not only in the sector where the improvement happens to take place, but also in other sectors which take up and modify the improvement still further.

Another way of presenting this is in terms of the ideas of natural trajectory and technological opportunity elaborated in Chapters 2 and 3. The 'New Technology System' of Freeman *et al.* can be seen as a set of powerful new natural trajectories created by some core advances in technology. During the long wave boom the rapid and diverse exploitation of these trajectories in a variety of fields lifts the upper limits for best practice in many industrial processes, and the ceilings for performance and specification in products and services. For the post war boom of the 1950s and 1960s they present substantial case study material to support the view that this role was played by two such technologies; those of synthetic materials, and 'first generation' electronics. (Freeman *et al.*, 1982, Chs 5 and 6). Both of these technical areas clearly also satisfy the requirements laid down by Rosenberg and Frischtak (1983) for a 'T' cluster in their alternative account of innovation clustering. They distinguish between 'T' clusters of innovations which are technically related in some way, and 'M' clusters which are not technically related but cluster because of the common stimulus of a generalised increase in demand, or other favourable macroeconomic conditions. Subsequently, Freeman (1984) has developed the notion of NTS using the work of Perez (1983). They argue that the new technologies can form a paradigm or 'style' which acquires the institutional accompaniments of training programmes, cadres of experienced personnel, received wisdom in areas of engineering practice and so on; thus facilitating the generalisation of the technologies. This raises the related question of whether these institutional structures, and other broader ones relating to financial, managerial and social contexts, can change at the same speed as technical possibilities when older paradigms or styles approach diminishing returns. This is an important issue which will be discussed further late in this chapter.

An important feature shared by Freeman *et al.*'s 'NTS' and Rosenberg and Frischtak's 'T' cluster is that they are more likely to occur, and to have significant effects, if they are located in 'infrastructure' industries such as transport, energy or communications. The capacity of new technologies in infrastructural roles to have abnormally high 'multiplier' effects on the growth potential of other industries and sectors is clear from contemporary discussions of information technology, as well as from analyses of railways and electrification in previous historical periods.

To turn now to the second essential feature of a new technology system, it is not only important that the specific technologies be powerful, but that their effects stem from the clustering of their *diffusion processes* rather than clustering of the *innovations*. Freeman *et al.* point out that the innovations themselves, being necessarily singular, can have but a limited economic impact on other firms and industries. It is in their diffusion that these impacts arise and this, they argue, is the sense in which Schumpeter wrote and which they wish to continue. This argument clearly differentiates their position from that of Mensch.

It should be noted however, that there are strong reasons to expect some connections between the clustering of diffusion processes and the clustering of technically *related* innovations. In Chapter 5 it has been argued (following Metcalfe, 1982) that the changing incentives to adopters and suppliers during the diffusion process flow in no small part from the technically related innovations which follow the initial innovation and continually improve its performance or lower its cost. This argument provides a link between the NTS concept and the 'T' cluster used in Rosenberg and Frischtak's analysis. Despite their general criticism of the long wave hypothesis this point in their paper seems more to support than to undermine the long wave theory. Furthermore, in a recent careful study of the clustering of innovations, Kleinknecht (1984) has used a wider range of empirical sources and a more subtle approach which considerably modifies the earlier clustering hypothesis. He argues that the clusters of radical product innovations are actually 'waves of innovation' which straddle the late depression/ early boom period and have some degree of technical and sectoral relatedness. This position is very much closer to the NTS concept of Freeman *et al.* and to the 'T' cluster of Rosenberg and Frischtak.

Combining some of the ideas from the preceding discussion; if diffusion processes and their related secondary innovations embody new technological trajectories which have intrinsically greater than average capacity for performance improvement, and for learning and transfer from other products also using the technology, then these diffusion processes will create greater than average benefits in terms of productivity, market growth, profit and all the other vital signs of a new sector's growth. The improvements in active electronic components and circuits, and their rapidly multiplying use in various consumer durables during the post-war boom is a clear example of this phenomenon. Similar examples from the synthetic materials sector can also be documented.

7.4   THE ROLE OF THE LABOUR MARKET

If the NTS plays the role of engine in the Freeman theory of long waves, then the role of transmission is played to a large extent by the functioning of the labour market under the impact of the diffusion of the NTS. Before turning to Freeman's account recall first the conventional approaches to the effect of technological changes on employment discussed in section 6.7. It is particularly appropriate to make this detour at a time when the debates about the possibility of long term unemployment being 'unavoidable' have been revived by the presence of a new impetus to automation from information technology and robotics.

As has been pointed out in Chapter 6, no economist has ever disputed that technical changes displace particular jobs in particular industries at particular times. However, approaches based on the efficient functioning of markets for goods, capital and labour have suggested that compensation mechanisms would, *other things being equal*, create new job opportunities in other industries and places. According to this view, technologically caused productivity growth will generate extra profits, increases in wages, or reductions in the relative prices of the goods produced. The first and second of these create new demand directly (either for capital goods or for consumer goods) and the third creates new demand indirectly by freeing some income for other purposes, or simply by increasing the level of consumption of the cheapened good. Since all of this demand must be supplied by new output, resources from the increment of growth in output will be invested in capital and in new jobs to create that new output. The only logical limit to the process is the scope of wants and needs of the population; the only constraints on the rate of growth are, as was shown in sections 6.3 and 6.6, the rate of growth of the population, the rate of technical change, and the proportion of output consumed as compared to the proportion invested in the resources to create future output.

But this is a simplified analysis, and the key phrase which enables it to be consistent at this level of abstraction is 'other things being equal'; but of course they are not. There are considerable obstacles to the easy movement of displaced labour from one set of activities to another resulting from skill patterns, lack of information, geography and social, cultural and many other factors. Many of these obstacles also apply to the mobility of capital. These are the factors frequently adduced to explain regional differences in employment levels. More fundamentally, as was shown in the discussion of Pasinetti's work in Chapter 6, if rates of technical change and demand growth vary across industries, there is no necessity for compensation mechanisms to be 'exact', and so net unemployment or labour shortages could result.

We conclude therefore, that if technical change is a major force for structural change, and economic growth is in a sense a consequence of structural change, then the analysis of employment change must be more directly related to the analysis of structural change, rather than being treated as compensation processes at an aggregate level. Furthermore, if long waves are as long as they appear to be, then the obstacles to compensation must take a very powerful form to explain those prolonged periods of unemployment and underemployment which do arise during long wave downswings. From the point of view of the labour economist then, the requirement is for a long run theory of the level of employment and its relationship to the diffusion of the NTS and the phase of the long wave. This is quite a different style of analysis from the short run analysis of imperfections, and from the cyclical underutilisation of resources in the short business cycle.

Freeman *et al.* offer a model of this long run character as a central part of their long wave theory. The principal features of it can be seen in Table 7.4, which also serves to summarise the stages of evolution of the New Technology System. They argue that throughout the *preceding* long wave, both upswing and downswing, the basic inventions and innovations which will form the basis of the New Technology System of the *next* long wave may emerge on a small scale either in new small firms or in branches of larger firms. Their economic and employment significance will be negligible in the context of the aggregate economy, but they will act as a new centre of gravity for the development of specialist skills both in technology and associated management. Once the next long wave upswing does get under way (the precise mechanism of the lower turning point is not explained in Freeman's account and will be dealt with below), the new sectors of industry may generate substantial new employment opportunities and also stimulate employment growth in existing industries upstream and downstream of the new industries. Furthermore, the new skills used in some of the jobs in the new industries will be in short supply and therefore the wage rates in those occupations will tend to rise. New training programmes may be set up to overcome these skill bottlenecks but they tend to be slower in appearing than is required, thus permitting some shortage to continue.

These pay rises act as a benchmark in the general setting of wages through comparability claims and, given that the whole labour market is tending toward full employment because of the prosperity of the boom, there is a powerful upward pressure on wages in general. At the upper turning point of the long wave these pressures begin to combine in an unpleasant manner with the incipient slowdown of the growth of the new sectors. There are two aspects to this problem. Firstly: if the rates of output growth are not maintained, due to market saturation, diminishing returns to technological trajectories and exhausted economies of scale, yet upward pressure on wages continues as a result of institutional factors, then it is likely that inflation will be the result. (In fact there is inflationary pressure originating in the capital market also see Freeman *et al.*, p. 190). In the long run this inflation has negative effects on investment both through corporate policies and through state policies.

Secondly, the technologies of the new sectors will, by this time, have been refined, together with other standard technologies, and become more capital-intensive. Thus their labour displacing role will overshadow their job-creating role. Electronics for example, created many new jobs in the 1950s and 1960s as new production sites were set up and as the early phases of growth of the electronic goods industries employed labour-intensive production techniques. But as the industries and the technologies matured, they became more and more effective as displacers of labour in both their own and other industries. (See Stoneman, 1976 for the example of computers).

As the depression continues in this model, investment tends to be more and more directed towards rationalisation rather than expansion against a background of intensified competition. The general picture then, is of unemployment rising as a result of both 'conventional' recessionary features of the economy and structural features related to the bias of technology. To the extent that unemployment does rise it has negative effects on aggregate demand and exacerbates the situation. The transition from virtuous circle to vicious circle is complete as a result of the interaction, over a long time period, of the 'New Technology System' with its markets, and with the labour market.

This model of the origins, diffusion and effects of a New Technology System incorporates much that is useful from existing literature on technical change and industrial and macroeconomics, while nevertheless going beyond the latter's normal short run

TABLE 7.4 *A simplified schematic representation of new technological systems*

	Previous Kondratiev	Recovery and boom	'Main carrier' Kondratiev		Depression
			Stagflation		
Research invention	Basic inventions and basic science coupled to technical exploitation. Key patents, many prototypes. Early basic innovations	Intensive applied R & D for new products and applications, and for back-up to trouble shooting from production experience. Families of related basic innovations.	Continuing high levels of research and inventive activity with emphasis shifting to cost-saving. Basic process as well as improvement inventions are sought.		R & D-investment becomes less attractive. Despite the fact that firms try to maintain their level of research it becomes increasingly difficult to do so with the slackening of their sales. At the same time the volume of sales required to amortize the cost of R & D is steadily increasing. Basic process innovations still attractive to management but may meet with social resistance.
Design	Imaginative leaps. Rapid changes. No standardization, competing design philosophies. Some disasters.	Still big new developments but increasing role of standardization and regulation.	Technical change still rapid but increasing emphasis on cost and standard components.		Routine 'model' type changes and minor improvements of cumulative importance.

Production	One-off experimental and moving to small batch. Close link with R & D and design. Negligible scale economies.	Move to larger batches and where applicable flow processes and mass production. Economies of scale begin to be important.	Major economies of scale affecting labour and capital but especially labour. Larger firms.	The slowdown in output and productivity growth leads to over-production and excess capacity in some of the modern industries. These structural problems are 'cumulative and self-reinforcing', with repercussions for the economy at large, and lead to a further decline of economic activity.
Investment	High risk speculative, small scale. Some inventor–entrepreneurs. Some large firms. Fairly labour-intensive. Problems of venture capital.	Bunching up heavy investment in build-up of new capacity. Band-wagon effects. Large and small firms attracted by high profits and new opportunities.	Initially continuing heavy investment but shifting to rationalisation. Continuing rapid growth, but increasingly large sums required to finace R & D and rising capital costs. Rising capital intensity.	Relatively low levels of investment. Underutilization of the capital stock in some of the most modern sectors of the economy; low profit margins and the general 'pessimistic mood' with regard to expectations lead entrepreneurs to be very (over) cautious in relation to new

(Continued on page 188)

(Table 7.4 continued)

| | Previous Kondratiev | 'Main carrier' Kondratiev | | |
		Recovery and boom	Stagflation	Depression
				investment opportunities. Investment which will take place will be primarily directed towards rationalisation. Search for new investment opportunities abroad.
Market structure and demand	Innovator monopolies. Strong consumer resistance and ignorance. Some new small firms to promote basic innovations.	Intense technological competition for better design and performance. Falling prices. Big fashion effects. Many new entrants in early build-up.	Growing concentration. Intense technological competition and some price competition. Strong pressure to export and exploit scale economies.	Even stronger trend to oligopoly or monopoly structure. Bankruptcies and mergers.
Labour	Small-scale employment-generating effects. High proportion of skilled labour, engineers and technicians. Training and	Major employment-generating effects as production expands. New training and education facilities set up and expand	Employment growth slows down, and as capital intensity rises, some jobs become increasingly routine.	Employment growth comes to a halt. Unemployment rising. In addition to the continuing labour displacement effects of

learning on the job and in R & D.

rapidly. New skills in short supply. Rapid increase in pay.

rationalisation investments, employment suffers (in the first instance) from the general recessional and depressional tendencies in the economy at large.

| Employment effects on other industries and services. | Negligible, but imaginative engineers, managers and inventors are thinking about them and planning and investing accordingly. | Substantial secondary effects, mainly employment generating but gradually swinging to displacement. | Labour displacement effects, as new technology now firmly established and strongly cost-reducing. | Continuing labour displacement as new technology penetrates remaining industries and services. |

Source: Freeman, Clark and Soete (1982)

focus. There is however, a gap in the theory in its explanation of the lower turning point and the factors which shift investment from the old technical pathways to the New Technology System. This was the role of the depression trigger in Mensch's work, but we have seen that his is not a sustainable explanation. If we are not to use a depression trigger mechanism to explain the lower turning point what can be put in its place? To answer this question it is necessary to consider in more detail the interaction of the technical and economic variables with their institutional context.

7.5   INSTITUTIONAL INNOVATION IN LONG WAVES

On both theoretical and historical grounds we can expect an economy in the grip of a long wave depression to experience all manner of social, institutional and political stresses and strains. Many of the important institutions of industrial life such as managerial structures, cultures and skills; training programmes; financial practices; labour market structures and government policies toward industry may adapt their shape to adjust to the specific features of the New Technology System in the long wave boom. As the New Technology System loses its ability to fuel economic growth, and the first signs of the next New Technology System emerge, it can become clear that there is a growing mis-match between the old institutions and the attitudes they imply, and the characteristics of the next New Technology System. Added to this there may be generic conflicts over income distribution, over regional and inter-industrial allocations of resources, and over the distribution of the costs of recession. There is no reason at all to suppose that the processes of change and adaptation in these social and political spheres move with the same rhythm as those in the narrowly economic and technical spheres. In fact they are likely to be slower and more unpredictable, depending as they do, not so much on stable relationships between simple variables, but on complex and contingent historical events, intrinsically unique and irreversible. Furthermore, when we relax the assumption of an abstract single economy, and acknowledge the reality of interacting national economies with military and political interests; cut across by increasingly complex business organisations which do not have a simple or stable relationship to these national territories, then it becomes clear that there may be substantial variation in the way in which institutional change occurs under the twin pressures of the recession, and the possibilities of the new New Technology System.

A number of authors have drawn attention in different ways to this political and contingent character of the lower turning point. Dosi (1983) has offered the general observation that long waves may be seen as periods of increasing and diminishing compatibility of economic and institutional structures. At the lower turning point the direction and rate of institutional change becomes a key limiting factor on economic growth. Perez (1983) has written in similar terms of the boom period of the long wave as one of a good 'match' between the technological paradigm and the institutional framework, and the depression as a period of mis-match. Gordon *et al.* (1982) have argued, and attempted to document for the case of the US, that for each long wave there is a characteristic 'social structure of accumulation' which results from key institutional innovations in the spheres of financial markets, labour relations, government patterns of intervention, international trading relations and cultural determinants of consumption patterns. These arguments share an emphasis on the institutional framework of the economic system and its long-run evolution, rather than the short-run equilibrium characteristics of an economy whose structure is 'abstracted away'.

The contributions of these authors therefore suggest that institutional change is the other necessary condition, along with new technologies capable of generating growth nodes in the economy, for the movement from long wave depression to boom. But there is still the question remaining of how the transition is accomplished. To a large extent this must be a question for empirical investigation in the case of specific long waves and countries, and not a matter of theoretical decrees. Gordon *et al.* (1982) have attempted this for the US; much more work on other countries, on their managerial systems, labour markets, and on international political and economic relations, is necessary before the usefulness of this approach can be finally decided. However, one author deserves mention for illuminating one dimension of how the technical and institutional changes may be linked together in a large number of cases. Mandel (1975) argues that the underlying variable which must explain the movement from recession to boom is investors' long term *expectations* of profit rates. If these begin to move in a favourable direction then the rate of investment can begin to pick up, utilise the new opportunities, and nudge the economy back toward a virtuous circle of productivity growth and demand growth. This may seem an obvious line of argument, but it is important to register that long term expectations of profit are always a political phenomenon as well as a technical one. No matter how clear the superiority of the new technologies and their long term strategic significance, the stability and acceptability of the social and political framework of market-based economic activity is always a major influence on the rate and direction of investment.

Mandel's own interpretation of this analysis is to argue that the lower turning points have normally involved a historic shift in the balance of forces between the working class and the owners and controllers of industrial capital. He argues this for example, in the case of the effects of Fascism and war on the European workers in the 1930s and 1940s, together with the changed position of the US in relation to the other major powers. (Mandel, 1975, Ch. 6). There is much to recommend Mandel's analysis but it is surely too simplified to see the lower turning points only as 'defeats' for the working class. The thrust of the analysis of the other authors mentioned, which seems to us more convincing, is to see the institutional innovations of each turning point as the establishment of a new package of compromises between a variety of groups, whose interests may diverge and coincide along a variety of dimensions, both intra- and inter-nationally. What matters from the point of view of the long wave model, is that new ground rules are established in the social and political order, which are compatible with the growth of the new industries, and which will be relatively enduring. It is clear that, except in the case of dictatorships, this latter requirement of endurance must mean that they confer real benefits which are perceived as acceptable, to most of the major interest groups in society. Without elaborating further, it will readily be seen that many historians' interpretations of the 'social democratic consensus', and the creation of state guarantees of social standards which characterised the political landscape of immediate post war Europe and America, can add some support to the analysis put forward here (Gough, 1979).

7.6 INFORMATION TECHNOLOGY AND THE CURRENT LONG WAVE

The contemporary coincidence of debates over 'new technologies' and institutional changes in industrial societies is currently centred on information technology. These familiar debates concerning availability of skills, future relationships between work and leisure, technical stimuli to economic revival in depressed countries and regions, patterns of control and demarcation in industrial relations and labour markets, and their relationship to the pressures of 'new technology' are cast into a broader context by seeing them as part of the lower turning point of the current long wave. In particular, the interaction of market and non-market forms of control on technical progress is raised in a particularly sharp manner. These issues are discussed further in Part III of this book.

The possibility that information technology will be the core of the next NTS' is relevant to the discussion of two further dimensions to the relationship between technical change and long waves. These concern the direction of process innovation over the course of the last three long waves, and the fact that the NTS may require a 'capital cheapening' character as well as the ability to generate new product areas. Taking the first of these points, Coombs has argued (1984a, 1984b, 1985) that mechanisation of industrial processes has followed a structured path of evolution since the mid-nineteenth century which is related in an interesting way to the long wave hypothesis.

In brief, Coombs argues that the focus of mechanisation in leading industrial sectors, embodied in the most advanced products of the capital goods sectors, has progressively shifted from transformation functions, to transfer functions and finally to control functions, and that in each case the transition has had some common features. As the productivity gains from mechanisation of one function have begun to encounter diminishing returns, a bottleneck has emerged which serves to focus attention on the associated function. These bottlenecks have been shown to be coincident with the downswings of long waves and have contributed to the difficulty of maintaining previous productivity growth rates. A feature of the lower turning point has therefore been the early use of some elements of the emerging new technologies in process innovations. These innovations certainly contribute to altering the division of labour and industrial relations practices in industry, and may contribute to elevating business expectations concerning the broader application of the new technologies.

Following the initial mechanisation of a substantial number of control functions in industrial processes in the post-war boom, a new bottleneck is now emerging. This lies in the fact that existing control mechanisation is very inflexible with respect to variation in batch size, product specification and product mix. This is increasingly a problem as mature industries seek to increase their degree of product differentiation, and as the focus of automation extends to industries and activities which have hitherto been difficult to mechanise precisely because their activities are so variable (for example, small batch engineering, office work, service sectors). Information technology, with its emphasis on reprogrammability, offers the potential for a new flexible mechanisation of control functions in industrial engineering which can make dramatic improvements in best-practice levels of productivity, and extend capital-intensive techniques to areas of industry and commerce previously notorious for their slow rates of technical and organisational change. This is a dimension of technical change in long waves absent from other accounts, but entirely complementary to the emphasis on the generation of new product areas in Freeman's model. These arguments are discussed at length in Blackburn, Coombs and Green (1985).

The second point concerns capital cheapening. Soete and Dosi (1983) have argued from substantial evidence, that some areas of the electronics industry which are involved in the information technology NTS have recently been unique amongst industrial sectors in showing substantial improvements in capital productivity. Such a phenomenon is, in their view, the result of a rapid rate of increase in technical performance combined with a rapid lowering of unit costs as a result of learning by doing, economies of scale, and widening of markets. If such powerful attributes are present in a technology which can then find broader applicability, their diffusion may bring the dual benefits (from the point of view of growth stimulation) of raising expectations of future profit levels, *and* generating more employment per increment of investment. This latter point would be a dramatic reversal of the trends in the OECD area over the past two decades, with a very powerful influence on demand stimulation.

It is interesting to note that one reason for the increases in capital productivity identified by Soete and Dosi may be the increase in production flexibility identified by Coombs. A second reason may be the penetration of the technology into the hitherto forbidden areas of application such as small batch production and service industries; where the low base levels of productivity growth make dramatic improvements all the more possible.

7.7 CONCLUSION

The first part of this chapter discussed the 'productivity puzzle' in the context of Pasinetti's analysis of structural change in which sectoral rates of technical change and demand growth contribute to fluctuations in aggregate growth and employment. The subsequent discussion has examined the arguments for some non-random character to attach to these fluctuations as a result of some clustering of technical change: either the innovations, the diffusion processes, or both. Our conclusion has been that there are some plausible mechanisms to explain such clustering in the case of diffusion processes. These mechanisms are both economic, resulting from the tendency of industries to exploit technical paradigms to maturity before switching to new ones, and institutional, resulting from the difficulty of altering institutional frameworks with the same speed as technical possibilities. In essence, these arguments make a case for recognising that technical change over the very long run is not homogeneous, and that there are some major changes in technological infrastructure which are of greater significance for long term growth.

The changes in the relative sizes and connections between different sectors of industry brought about by these infrastructural changes are so substantial that they create powerful channels within which the short run economic processes of economics text-books operate. But these channels are not only market structures, they are also the overarching structures of institutions and social processes. The evolution of the industrial and social system from one set of channels to another is the very essence of economic *development*; a much broader concept than economic *growth*. The long wave analysis; seen not as a deterministic cyclical model, but rather as an explanation of varying periods of alternating growth and structural crisis, is a useful frame-work for understanding some of the broad features of structural change and for the conduct of further empirical and theoretical research. Of vital importance is the realisation that the institutional changes related to long waves are the very stuff of politics. The subtly changing frontier between Adam Smith's invisible hand, and Chandler's visible hand is nowhere more apparent than in the long run analysis of technical and structural change.

Part III
Political and Social Aspects of Technological Change

8 Government Intervention in Technical Change

8.1 INTRODUCTION

Earlier chapters have discussed the complex and interactive nature of the relationship between technological and economic changes at the macroeconomic level; of the relationship between market signals, technological discoveries and the decision-making process at the microeconomic level of the firm; of the relationship between micro-level and macro-level events; and of the changing nature of these relationships over time.

As pointed out previously, a high degree of coordination of individual activities is required if a society is to exhibit the minimum degree of stability necessary to survive. The institutions examined so far, the firm and the market, exert this coordinating function in a number of ways, but no one would expect them to be able to coordinate all human activities. For example, the introduction and application of new laws is normally considered the responsibility of national or local governments. Thus there are spheres of human activities in which coordination is considered a natural function of national governments. However, in the sphere of human activity with which this book is mostly concerned, the economic sphere, this division of labour in terms of coordination is far less clear. Many have argued in the past, and argue nowadays, that the market or the market plus the firm, constitute a self-regulating system and therefore that state intervention is either unjustified or possibly dangerous. It is the technological dimension of this apparently unending debate which constitutes the main subject of this chapter and of the following one.

The primary concern of government is macroeconomic policy, to influence the overall level of economic activity and consequently the growth of output, the level of employment, the rate of inflation and so on (Pickering and Jones, 1984). The development of an industrial policy cannot be isolated from economic policy in general (or from other policies, such as social or manpower policies). A great many facets of government policy will affect industry, in addition to the particular policies specifically aimed at influencing the growth and activities of industry. In turn, even without any government policy aimed at the promotion or regulation of technical change as such, a government's industrial policy will have an effect on technical change.

Industry is influenced by a vast array of laws, regulations and voluntary agreements on such matters as tax, employment of children, pensions, national insurance, prices, exports, advertising, degree of monopolisation, disabled employees, sex or race discrimination, design and location of factories and offices, availability of capital, terms and physical conditions of employment and training of personnel (Teeling-Smith, 1969; Horwitz, 1979), only some of which are the result of a specific industrial policy. Some of these laws, regulations, and agreements will in turn, directly or indirectly, intentionally or unintentionally, influence the nature and rate of technological change. For example, increases or reductions in tax may influence the amount spent on R & D or capital investment in new machinery; alternatively and more directly, tax incentives or credits may be used to offset R & D expenses specifically, as has been attempted in Canada (Hollomon, 1979) and recently in the US (Joyce, 1985).

This chapter will be mainly concerned with government activities with the *primary* goal of influencing the rate, direction or consequences of technological change. It will be seen, however, that since the majority of R & D and almost all technological innovations are carried out by industry, the overlap between industry policy and technology policy is considerable, and the justification for government finance of R & D is very similar to that of direct government intervention in industry, (Horwitz, 1979; Pickering and Jones, 1984) while government regulation of the unwanted consequences of technical change is almost always synonymous with regulation of industry. The analysis of actual postwar government policies for promotion and regulation of technology in Chapter 9 will thus be in practice a discussion of *industrial* as well as specifically *technology* policy. These policies will be discussed with reference to different political parties, different examples of technical change or its consequences, and different countries – although we shall concentrate primarily on the UK.

In this chapter, we shall discuss the reasons why government has chosen to intervene in the process of technological change. The previous chapters have detailed many forces derived from the operations of firms and markets which might be thought adequate to establish and to pursue those areas of science, technology and innovation which best meet demand. Nevertheless, even the present governments of the industrial economies of the 1980s, which show a general increase in commitment to the importance of market forces and competition, are simultaneously committed to a complicated mass of legal and policy instruments for the promotion and control of technological (and industrial) change. Governments have attempted in some sectors to replace direct support for innovation with policies designed to influence the ability and willingness of private firms to innovate ('by enhancing their ability to do so, increasing their desire to do so or increasing their fear of not doing so.' Horwitz, 1979). But this amounts to an equally complex set of policies for influencing technological change: only some policies of direct intervention have been replaced by indirect intervention.

Even at the height of *laissez-faire* ideology in Victorian Britain, the government had a conscious policy of state intervention. *Laissez-faire* was itself the product of deliberate state action – 'continuous, centrally organised and controlled interventionism' and 'Administrators had to be continually on the watch to ensure the free working of the system' (Polanyi, 1957). And 'Britain's position as a *laissez-faire* industrial nation was developed by imposing trading relations upon other countries by the use of military force' (Hodgson, 1984). The British government spent about ten per cent of GNP in maintaining 'free trade' and an environment in which industry was able to operate as it saw fit. Similarly today, in the interests of competition and the 'free market', 'monetarist' governments employ anti-trust legislation and a range of other policy instruments to prevent the natural tendency of the market economy to increasing concentration and monopolisation.

Within the *laissez-faire* environment maintained by the state in nineteenth-century Britain, technological change and industry were the business of inventors and entrepreneurs, and not of the government. Even the relatively indirect action of changing the patent laws to encourage British innovators was resisted for many

years. Other countries however, notably Germany after 1871 and
subsequently the United States, industrialised later than Britain
and on the basis of a great deal more direct government assistance
than was considered proper in Britain. This chapter analyses the
historical development of government policy from *laissez-faire* to
the considerable degree of direct intervention existing today.

8.2   THE POSSIBILITY OF INTERVENTION

It has been argued at times that intervention is impossible because
science is unpredictable and therefore essentially unplannable.
Why do we not give scientists whatever money they need to do the
research they want to do? Some of it would be bound to make an
important economic or intellectual contribution to society.

Even before the Second World War, when government intervention in science and technology policy and industrial planning
was far less substantial than it is today, this view though often
expressed, was easily countered by reference to the basic facts of
the world of science. Most people actually engaged in science and
technology are working in fields which, by their very nature, are
already highly controlled and planned. Industrial science and
technology is not controlled and planned by scientists and technologists alone but by the managerial structure of industry which, as
we have shown in Chapters 2 and 3, is concerned with costs,
profits, growth and other individual company goals rather than the
logic and momentum of science itself. Some R & D areas are
controlled by national considerations of strategy, prestige or defence.

Even academic research is not free from constraints. For a large
part of scientists' careers they are likely to be working on projects
planned and organised by other people. Even senior academics
rely on funds from outside bodies, such as the research councils,
which are awarded on the basis of a refereeing system, and criteria
including what is and is not desirable for government bodies to
fund. In 1969 a government publication proposed a model for
quantifying the economic benefits of basic or 'curiosity-oriented'
scientific research (Byatt and Cohen, 1969). Although their approach was subsequently criticised, the notion of economic justification of academic research has become an explicit consideration.

The Rothschild Report (CPRS, 1971) more formally introduced the idea of customer-contractor relationships and the long term economic justification of academic research into science policy in Britain (Gummett, 1983). Following the 1982 Rothschild report on the Social Science Research Council (Rothschild, 1982), the Research Councils were criticised for their support of 'esoteric research'.

Scientific criteria, including the soundness of the proposed methodology or the expected contribution of the research to the understanding of certain phenomena (as perceived by the decision-making committees) play a decisive part in the selection of projects for funding. But reference to non-scientific criteria, such as the probable social or economic value of the research or even its compatibility with prevailing government ideology, are likely also to be used in the selection process. In the choice between alternative projects, actual priorities must be established, even if they are implicit rather than explicit.

The decision-making committees are composed of senior academics, who might be seen as engaging in the self-regulating process described by early sociologists of science (for example, Merton, 1942). To some extent they are, but the links between these scientists and industrial and government establishments mean that their views about what is good science are related to, if not totally determined by, 'national priorities'. The tension between these two influences can be seen in government reports, such as the Rothschild reports already mentioned, which have tried to shift the balance still further, and may be seen as attempts to make non-scientific criteria and priorities rather more explicit.

Even during the period of maximum growth in spending on science and technology 1950–65, between 40 per cent and 70 per cent of applications to UK and US government funding bodies were unsuccessful because funds and personnel were not available in unlimited supply. Both scientific and non-scientific criteria were used to select projects for receipt of funds (Jevons, 1973).

Choices must always be made, and hence priorities established. The question thus becomes, not should science be controlled, but how can it be controlled or planned most effectively? And according to what criteria and in whose interests? Priorities are usually established either by the operation of market mechanisms, decisions made by managers of private industry in response to signals

Political and Social Aspects

TABLE 8.1   *Resources devoted to R & D by sector of performance*

	USA 1985	% UK 1983	Japan 1983
Business Enterprise	71.2%	61.0%	63.5%
Government	15.6%	25.1%	13.5%
Other	13.2%	13.8%	23.0%
Total	100%	100%*	100%

* does not add up to 100% as a result of rounding.

Source: OECD (1985)

TABLE 8.2   *Resources devoted to R & D by source of funds*

	USA 1985	% UK 1983	Japan 1983
Business Enterprise	48.9%	42.1%	65.2%
Government	49.2%	50.2%	24.0%
Other	1.9%	7.7%	10.8%
Total	100%	100%	100%

Source: OECD (1985)

or anticipated signals from the market, or by government policy, or by some combination. Chapter 10 will consider influences other than those of government or industry on technological change.

In advanced capitalist societies the interplay of markets and business decisions determine the nature of the goods that are produced and the methods for producing them, and consequently the rate and direction of technical change, though we have seen in earlier chapters that the role of demand is not dominant in any simple sense. Table 8.1 shows that approximately two thirds of R & D is done by industry in Britain and the US. Industry pays for nearly half of the R & D done in both countries, see Table 8.2. We have seen in Chapter 2 that the corporate policies of companies are formulated in an attempt to respond to market signals, the perceived potential of technologies, and strategic interests of firms. Consequently resources are allocated to *certain areas* of research, development and design. Technological change is therefore never 'automatic', but is always the resultant of many attempts to 'intervene'.

8.3   WHY DOES GOVERNMENT INTERVENE?

It is clear that government policies would affect the rate and direction of technological change as a consequence of industrial and other policies, even without the adoption of policies aimed specifically at the promotion or regulation of technological change. We now turn to the question of why governments do attempt to influence technical change specifically, in some cases intervening directly by sponsoring many areas of scientific and technological research and development, and in other cases by promoting the adoption of innovations in various ways.

The justification for government intervention in the process of innovation in Britain has always been based by both Labour and Conservative governments on an acceptance of privately owned industry and the market mechanism. Government science and technology policies have been intended as a *modification* to the activities of private industry, not as a *substitute* for them. Their rationale is to benefit industry as a whole (as well as 'the general public'), even though some individual firms may not necessarily benefit in the short term; and even though some industrialists may oppose some of the policies.

Just as Chadwick's campaign for public health led to a more efficient work force, and the Alkali Acts stimulated cost-reducing as well as pollution-reducing innovation, as described more fully later in this chapter, so the policies of all post-Second World War UK governments – including nationalisation – have been aimed at making privately owned industry as a whole and the market mechanism work better, not at replacing it with something else. (A full discussion of the economic rationale for public ownership may be found in Jones and Cockerill, 1984). Whether or not the policies have been successful, of course, is another question; and one that we intend to try and answer in this chapter, and in Chapter 9.

Government's own justification for intervention is that the market mechanism alone fails or does not adequately ensure the optimum allocation of resources for the benefit of society, for industry in general or even for the individual enterprise itself. The argument of market failure is one that is used as the economic justification for government intervention in industry generally (Pickering and Jones, 1984). One of the areas where market failure arises is in the provision of public goods (Pickering and Jones, p. 313). Public goods tend to have one or both of two economic characteristics: non-rivalry in supply (the use of the good by an extra customer does not impose any extra cost), and non-excludability (the consumption by one individual does not reduce the amount available for consumption and it is difficult to exclude anyone from consumption) (Bator, 1958). Examples of public goods are roads, bridges, rail transport. defence and public health. They are usually provided by government or at least with government subsidy. Several areas of science and technology may also be considered 'public goods' using this definition, although imperfect information and patent law still allow many areas to remain private property.

The underlying rationale for government support of research and development is usually based, implicitly or explicitly, on theories relating science and technology to economic growth. During the post-war boom almost all governments of the advanced capitalist countries developed explicit policies for the fostering of economic growth as a means to pay for increased wages, health and welfare policies and defence (Maddison, 1964). The actual policies varied from country to country but usually included some combination of demand management, investment incentives and the promotion of science and technology. In Britain, macroeconomic policy was strongly related to ideas of demand management, while science and technology policy was justified with reference to 'science-push' theories of innovation and economic growth.

By the late 1960s, however, 'science-push' *policies* were being questioned by decision-makers in both government and industry while 'science-push' *theories* were being modified by ideas about 'demand-pull' in the study of innovation (see Chapter 5, and Walsh, 1984). By the early 1970s the 'demand-pull school had scored a victory on points' (Freeman, 1979) at least in the interpretations of the policy makers (Mowery and Rosenberg, 1979). The changing fortunes of 'science-push' and 'demand-pull' theories were thus reflected in science and technology policy in an early upsurge in spending on R & D followed (once the upper turning point of the economic upswing had been reached) by cut-backs justified with reference to the results of research on innovation. However, it would be wrong to suggest that the theories alone gave rise directly to the policies: the interaction was two-way. The recognition that high spending on R & D was not an infallible recipe for proportionate growth in GNP stimulated sponsorship of the innovation studies in the first place. This recognition, together with the changing economic conditions, also strongly influenced the interpretation of the results in favour of 'demand-pull' theories which, in turn, were used to justify cut-backs.

In addition to assumptions about the causal relationship between science, technology, innovation and economic growth, a number of other reasons have been used to justify government sponsorship and regulation of technical change.

First of all, the scale of capital investment or R & D investment required for industries based on new technology, especially 'high technology', has often been such that individual firms cannot raise the necessary funds or accept the high risk involved in the development of the new technology. In these circumstances governments have sponsored research and development and made capital available or guaranteed loans. In some circumstances, for example where public utilities required massive capitalisation, governments have taken whole industries, or major sectors of them, into public ownership, in the interests of the rest of industry.

Secondly, governments have also provided funds to support industry in the face of international competition; either to support sectors (such as computers and microelectronics in Britain) which for strategic reasons the government believes should be competitive, or to protect others, (like the UK textile industry) that are not.

Thirdly, there are many areas of activity of importance to industry or to society as a whole, such as energy, transportation or telecommunications, where an individual enterprise may not necessarily benefit from making an investment in technological change. Here too, governments have supplied funds.

Fourthly, basic knowledge is likely to be useful to industry in the long term. In some industries where technical change is rapid such as pharmaceuticals, firms may directly use knowledge generated in basic research in universities. But if the market mechanism were the sole arbiter in allocating resources to basic research, the length of time between investment in basic research and the economic benefit received from it would be so great as to result in a shift of resources away from basic to applied research (Shonfield, 1981).

Fifthly, there are many areas of basic academic research which cannot be said to result in discoveries that lead directly to technological advances. The market mechanism would allocate resources to these areas even less efficiently. The scientific knowledge produced in these areas provides an infrastructure which influences what industrial researchers, engineers and inventors do, their methods and the knowledge with which they start: but it is often the case that individual industrial investors are not able to appropriate all the benefits of basic research and training activities. This consequently adds to the tendency for the market mechanism to allocate inadequate investment (Shonfield, 1981).

Sixthly, in sectors with very small units such as agriculture, it is now generally accepted that the market alone does not generate all the technical change that is economically and socially desirable or necessary. Individual farms are too small to have the skills and financial resources necessary for R & D and related activities.

Seventhly, in some service areas such as health, it is now widely argued that access to and provision of health care should not be governed (or solely governed) by the market mechanism. The relative share of private and public contributions to health provision, including R & D, varies from country to country and between governments of different political parties. But it is generally accepted that governments pay a large share, while the market is dominant in particular areas of R & D such as pharmaceuticals and instruments and equipment. In Britain the government's share is about two thirds of total expenditure on medical research.

The eighth area is defence, by definition a topic for government policy. Individual firms may well approve of defence spending to protect their country, for example from invasion, and may additionally agree with military activities aimed at, say, protection of overseas markets or raw material sources. But the market mechanism would not allocate many resources to defence R & D compared to economically oriented R & D.

Any activity – like defence – that accounts for up to two per cent of GNP is clearly economically significant; but in the defence area economic considerations are not paramount in governing the size of investment and where it goes. The primary considerations are of security and prestige rather than economic ones. The same considerations apply to projects thought by government to enhance national prestige, whether or not they are directly related to defence, such as Concorde. National prestige and strategic considerations also played an important part in government support for Rolls Royce in developing the RB211 engine in the early 1970s (see section 9.4).

There are very big differences in amounts spent on defence R & D between countries. The US and UK spend half to two thirds of government resources for R & D on defence and space. Government spending on defence and space R & D in both countries has been a major determinant of the rate and direction of technical change in both countries. US Government defence contracts, for example, played a major part in IBM's growth. Following work on electronic control systems and computers for use in missiles and satellites, IBM diversified into the market for non-military applications of their innovations.

The above arguments are all specific elaborations of the failure of market and company-based mechanisms to provide adequate resources for the development of science and technology considered necessary at the aggregate level for the integrity of an advanced industrial society. They all relate to the likelihood of a 'social return' on R & D being significantly higher than 'private returns' to investing firms (cf. Mansfield *et al.*, 1977) and/or the provision of public goods and/or the risk to individual firms.

There is one further area where the market fails to direct technology, and that concerns the control of its unwanted effects, or 'externalities'. Governments are concerned with advancing the overall well-being of society. All innovations have costs and benefits; but some innovations provide benefits to one group of people and costs to a different group. Government therefore regulates technical change with a view to optimising the costs and benefits to society as a whole or intervening in conflicts between groups with different interests in the consequences – costs and benefits – of technical change. The US saw a spate of laws passed during the period 1965–75 which regulated unwanted effects of technology. Other advanced countries followed (see Table 8.3). The late 1960s and early 1970s were the years of protest in the US: about civil rights and Vietnam as well as the social costs of advanced technology

TABLE 8.3   *Some US and UK legislation regulating the effects of technology*

Year	US	UK
1959	Kefauver-Harris Amendment to Food, Drug & Cosmetic Act (1938)	
1965	Water Quality Act	
1966	National Traffic and Motor Vehicle Safety Act	
1967	Air Quality Act	
1968		Medicines Act
1968		Countryside Act
1968		Clean Air Act
1970	Clean Air Amendments	
1972	Federal Water Pollution Control Act Amendment	Deposit of Poisonous Waste Act
1972	Noise Control Acts	Road Traffic Act
1973		Nature Conservancy Council Act
1974		Health & Safety at Work (etc.) Act
1974		Control of Pollution Act
1974		Road Traffic Act
1975	Energy Policy and Conservancy Act	
1976	Toxic Substances Control Act	
1981		Industrial Disease (Notification) Act
1981		Wild Life and Countryside Act

Source: Reppy (1979); Kaufman (1984); McGinty (1979)

(Reppy, 1979). Concern about environmental pollution, depletion of scarce resources, health and safety at work, noise, product safety and the invasion of privacy all became political, as well as technological issues and most resulted in the introduction of regulations.

There are economic as well as social and political arguments for regulation; for example, it is cheaper in the long run to prevent occupational illness than for health services to treat it, and it is more cost effective for firms to employ healthy workers who do not need time off as a result of occupational ill-health, than to save money by not installing safer machinery. However, governments usually justify their intervention in this sphere in terms of social well-being.

In principle, the market could operate to regulate the unwanted effects of technology if, for example, the social cost of manufacturing could be translated into an economic cost, or cash value, to be levied like a tax on the manufacturer, and consequently added on to the cost of raw materials, wages, depreciation, energy and so on. This would then be passed on to the final consumer, so that the price paid for the benefit of having the product reflected the cost to society as a whole of producing it.

For example, the benefits of a healthy working environment are not immediately or directly felt by the firm, so it may feel no incentive to invest in it; while in the case of resources or amenities that are not directly owned – like the atmosphere – the market normally fails to put a price on the costs and benefits of using them. Some sort of market mechanism could be brought into operation by the kind of tax system described above. In practice, however, government regulation is usually cheaper and easier. For example, it is usually cheaper and easier to prevent pollution occurring than to clean up afterwards, and governments usually prefer to impose a regulation that limits the amount of toxic material allowed in the environment rather than calculate the costs of it and impose a tax.

In the case of product safety, the market mechanism will fail to operate if consumers are unable to make a fully informed choice. The public are frequently not in a position to judge the safety of complex, technologically advanced products like motor cars or electrical equipment because they lack expert knowledge. If the market mechanism were to regulate the technology, it would be necessary to provide sufficient information to everyone likely to buy a particular product so they can choose accordingly. It is usually cheaper and easier to introduce controls at the production stage so that products have to meet certain standards before reaching the market. Thus 'public decisions replace private ones to mitigate the economic effects of monopoly power'. (Reppy, 1979).

Since 1974 the slower economic growth in the US has been reflected in a slowing down of the regulatory activities of government. It has become much more common for economic arguments against regulation – concerning the cost to the enterprise and the inhibition of innovation – to be accepted as reasons for shelving or modifying regulations (see Rothwell & Walsh, 1979). There have been attempts to revise existing regulations in response to the increasingly strong anti-regulation lobby (see Chapter 9). Britain and other advanced countries have been under the same pressure and have followed similar trends, but a little later. The effect of technological change of most concern in the late 1970s and early 1980s seems to have been the effects of advanced (especially microelectronic) technology on the quantity and quality of employment. This has so far been a concern that has been restricted to trade union negotiation – with government as well as private employers – rather than government intervention and will be discussed in Chapter 10.

To summarise, we have analysed the various rationales adopted by governments – and by industry – for increased intervention by the former in the affairs of the latter, in order to promote technological change and control its unwanted effects. These are based essentially on the inadequacy of market forces in promoting or regulating certain areas of science and technology which may be desirable in the interests of society or the economy as a whole, in the long as well as the short term.

While there may be some degree of consensus, after the event, over the existing scope of government intervention, this is by no means the case with *proposed* intervention. Most of the existing, and in some cases well established and universally accepted, areas of government intervention, were only achieved after a great deal of controversy and in many cases some very bitter political battles.

In developing and proposing effective government and industry policy towards technical change for the future, it is very instructive to analyse the way in which such policies have evolved in the past. The following section discusses a brief history of government intervention in science and technology.

8.4 HISTORICAL BACKGROUND TO GOVERNMENT INTERVENTION

In Britain, during and after the industrial revolution, innovation was not considered to be a matter for direct government intervention. Indeed, neither were any other aspects of industrial activity. A policy of non-intervention was deliberate. Individualism, *laissez-faire* and self-help were the mottos of the day in Victorian Britain. It was generally believed, by government and manufacturer alike that the 'manufacturers and traders were the best judges of how the country's economic resources should be developed' (Levinstein, 1883). It was the job of government policy to maintain a 'free market' environment within which the market forces would shape industrial and technical change. Levinstein commented: 'It is difficult to get the House [of Commons] to consider any commercial question of national importance' (1883). This attitude contrasts with those held in other countries, especially Germany, where industrialisation was strongly promoted with the help of a variety of government policies aimed at more direct intervention.

Levinstein was a chemist and entrepreneur whose firm was one of the cluster of British Dye Manufacturers that contributed to the revolution in dye production in the 1850s. Traditional, imported, natural materials were replaced by coal-tar derivatives, and the dyers' art by the science of organic chemistry. The early synthetic dye industry developed into the organic chemicals industry, producing drugs, explosives, agrochemicals, plastics, detergents and synthetic rubber in due course, while Levinstein's Ltd became part of ICI. But the major innovative activities in these fields took place in Germany rather than Britain after 1870 for a variety of reasons largely attributable to British *laissez-faire* compared to Germany's active promotion of technology and industry. This is discussed further later in this chapter.

In Britain, the idea of Peel, Gladstone and successive prime ministers was 'cheap government': severe restrictions of numbers of civil servants and as little intervention as possible. The industrialists had achieved their idea of a 'minimum state' which put no restrictions in the way of buying cheap and selling dear; but this concentration of government intervention on a *laissez-faire* environment was eventually counter-productive from the manufacturers' point of view. Levinstein campaigned for years, for example, before the government even amended the patent law to protect British invention and manufacture, or provided scientific and technical education. Both patent law and the provision of technical education were government activities which influenced the rate and direction of innovation: but were far from the direct intervention of the governments of sixty to one hundred years later.

A dramatic but illustrative sample of the *laissez-faire* attitudes of the UK can be found in the mid-nineteenth-century history of public health legislation. Fear of cholera, the most dramatic of epidemic diseases in the nineteenth century, and fear of revolution, which was seen as originating in the same breeding ground (the slums), pushed through the Public Health Act in 1848 (Briggs, 1961). Campaigners for this legislation, notably Edwin Chadwick, were not motivated by humanitarian sentiments but by a cool calculation of economic and social benefits. He argued that conditions in the towns were wasteful of resources and produced a population 'short-lived, improvident, reckless and intemperate' – and likely to be revolutionary (Finer, 1952).

No one could say that Chadwick did not have the interests of the manufacturers at heart. But he was too bold too soon. He rejected the free trade doctrines of the Manchester School as narrow and inadequate whilst they were at the height of their influence. His dogged application of utilitarian rationality led him to conclude in anticipation of common modern day economic notions of 'public goods' and 'natural monopoly' that in such areas as water supply, cemeteries and public sanitation, competition between private enterprise was more expensive and less efficient than public monopoly. (Briggs, 1954). But once the fear of cholera had subsided, *laissez-faire* ideology ensured that the legislation lasted only ten years. 'We prefer', exulted *The Times*' leader writer when the Board of Health was wound up in 1858, 'to take our chances of cholera and the rest rather than be bullied into health' (*The Times* 1858). The social problems arising from further industrialisation eventually made direct state intervention unavoidable. But still, every increase in state involvement to tackle the problem of health, education, welfare, sanitation and the provision of utilities, met with resistance.

Campaigners like Chadwick may (in some senses) be seen as forerunners of the pressure groups that grew and flourished in the 1970s on such issues as pollution or consumers' rights (see Chapter 10).

Chadwick represented the longer term interests of the manufacturers, while many of his opponents were concerned only about their profits in the short term, and freedom from government interference.

The first successful and systematic application of a direct government policy towards science and technology was the regulation of the chemical industry through the Alkali Acts in 1863 and 1874. These were described as 'the first instance in Victorian contractualist society of the expenditure of public funds for scientific protection of private property' (McLeod, 1965). The success of the campaign to introduce regulation of pollution by the chemical industry was largely due to the power of those who suffered from the pollution: the landowners, whose representatives dominated the House of Lords, and who represented at least as powerful a 'pressure group' as any environmentalist group one hundred years later. The manufacturers attempted to defend themselves by arguing that legal action against them would mean 'the extinction of the alkali trade' and consequently would seriously affect the manufacture of soap, glass and textiles. This argument of the survival of an industry versus the control of unwanted effects of its technology is still common today, for example in the case of the asbestos industry.

When they realised legislation was inevitable, however, the manufacturers decided to cooperate as fully as possible in order 'to guide the threatening legislation into as harmless channels as possible' (McLeod, 1965) – again a tactic employed today. The original proposals were successfully whittled down on the grounds that stringent controls would inhibit the development of the industry.

When the Alkali Inspectorate was set up in 1864, it worked in close and friendly cooperation with industry, here too setting the tone for future patterns of regulation in Britain. As it turned out, the manufacturers welcomed suggestions from the inspectors for new and profitable ways of transforming hydrochloric acid into hypochlorite and commercial bleach for the textile industry, thus reducing waste as well as pollution. The scientific and technical advice, provided by the state through the inspectorate, speeded up progress in the chemical industry (Hardie, 1950). On the other hand, the modifications to the legislation successfully negotiated by the manufacturers, in the long term rebounded on them: more stringent legislation would have stimulated investment in a replacement for the old Leblanc soda process. The more modern Solvay process, which was more efficient in its use of materials, was widely introduced in Europe while British manufacturers continued to use the Leblanc plant, as it represented the accumulation of a great deal of past capital investment. The British manufacturers found their competitive position severely eroded after about 1880 (Reuben and Burstall, 1973).

In recent years the argument that government regulation inhibits innovation has had wide currency among industrialists, and in government too. However, as the Alkali Acts and soda manufacture indicated in the 1880s, regulation also has the power to stimulate innovation. A more recent example of this may be found in the manufacture of PVC. The campaign for lower levels of emission of vinyl chloride in the early 1970s was initially opposed by some chemical firms on the grounds that manufacture of PVC would become uncompetitive: but when new regulations were established, they stimulated process innovations that not only decreased the workforce's exposure to vinyl chloride but were also more efficient in use of raw materials and thus more cost-effective (Green, 1982). Two product innovations in PVC flooring were also stimulated by regulations governing the conductivity of hospital operating theatre floors (Seal, 1968) and regulations banning the use of asbestos fillers in domestic flooring (Walsh and Roy, 1983). Chapter 9 discusses more fully the recent history of government regulation of technology.

The 1863 Alkali Act therefore pioneered the direct intervention of government in the control of technology and the 1870s saw the beginning of a steadily growing stream of social legislation, including the establishment of the control of some of the worst side-effects of technical change as a government responsibility. The 1874 Alkali Act extended the provisions of the 1863 Act which was further developed by the 1881 Alkali (etc.) Works Regulation Act. A new Public Health Act appeared in 1872. A succession of Mines Acts (such as those in 1872 and 1911) made some dent in the alarming rise in the rate of occupational diseases and accidents suffered by miners which accompanied deeper mining and other innovations. In the area of product quality and safety, various regulations also appeared, such as the 1875 Act restricting the adulteration of food.

The idea that the functions of the state could be limited to the maintenance of a 'free market', the preservation of life and property and the enforcement of contract was increasingly made untenable by the growing complexity of industrial society. Theory lagged behind changes in society as is often the case. By the 1880s sections of the employers, and the government, realised the inadequacy of *laissez-faire* and the need for a new programme. Members of the new generation of businessmen (such as Joseph Chamberlain) recognised that, in the age of the joint-stock company, monopolisation and imperialist expansion, capitalism would. have to take positive action if it was to maintain its stability. Political democracy and trade unionism, themselves essential components of stability in a capitalist society, made it obligatory to tackle the problems of health, education, sanitation, pollution, welfare and the provision of utilities, although the regulations invariably lagged well behind the growth of the problems. Control of pollution under the Alkali Acts lagged behind industrial production, for example, while the success of regulations in controlling the worst effects of mining on the workforce was constantly being overtaken by new mining innovations. Understaffing in the inspectorates was a problem then, and remains one to this day.

Nevertheless, by the 1870s it was established that the control of some of the worst side-effects of technology was a task for the government. It was much longer, however, before it was accepted that it was within the realm of government activity to try and *promote* or *stimulate* technology by any means more direct than the opening up of markets and sources of raw materials, at least in Britain. In Germany, the promotion of technology by direct government action was established in the 1870s. Very little was done in Britain until the First World War brought home the fact that British industry had been so thoroughly overtaken by German industry in some sectors that Britain was dependent on Germany for imports of dyes for uniforms, drugs, explosives, rubber and other chemicals, magnetos for transport, tungsten for steel, zinc for smelting, range-finders for guns and other optical and precision engineering equipment (Rose and Rose, 1970).

Britain's lead in world production of synthetic dyes (and subsequently all organic chemicals) had been overtaken by Germany after 1870. By 1900 Germany had 90 per cent of the world market. The German government had taken a leading role in stimulating industrial production and their policies played a major role in the success of the chemical industry there. Policy instruments included state-backed development of a banking system that provided the necessary capital for industrialisation, transport concessions, preferential duty on certain raw materials, favourable amendment to the patent law and state support for scientific and technical education and research.

Despite their success, it is important to note that these were still examples of *indirect* intervention by government. The German government did not sponsor or carry out any R & D or innovation itself; its policies were aimed at increasing industry's ability and willingness to do so. But they were policies based on a vastly different philosophy from the *laissez-faire* of Victorian Britain.

While a series of new institutions for organic chemistry were being built in Germany, the British Government was refusing money to the only British equivalent, the Royal College of Chemistry (Beer, 1959). British patent law was not amended until 1907 and the British Government was not shocked into sponsoring research, technical and scientific education and the rationalisation of industry until the First World War.

It is therefore reasonable to argue that *laissez-faire* was responsible for the decline of the British dye industry. 'So long as the textile industries got all the dyes they needed at the right prices, who cared where they came from? Who cared if the technological initiative passed to Germany? It was looked on as a natural economic change' (Cliffe, 1963). Britain had an external trade greater than that of France, Germany and Italy combined and three or four times that of the US. Investors had the pick of the world's investment market, so with ample opportunity to put money where the risk was known, why, they might have asked, should they have put their money into businesses that, much less securely, depended on science and technology? (Beer, 1959).

Germany, on the other hand, industrialised late. Government and industrialists recognised that they could not easily compete with Britain on the basis of price. Local raw materials were worse in quantity and quality than in Britain, while overseas trade in raw materials and manufactured goods was dominated by Britain. The only solution seemed to be to compete by making better products on the basis of technological advance (Lilley, 1973; Sohn-Rethel, 1978). This was a highly successful strategy and, it has been argued (see Chapter 6), many British manufacturers still have not recognised the advantages of 'non-price' competition. Given the steady erosion over the century of British industry's other early advantages (Walker, 1980), this lack of 'non-price' competitiveness is now a major weakness.

Another major difference between Britain throughout the nineteenth century and post-Bismarck Germany was the organisation of R & D. Bayer, Hoechst and BASF, the German chemical firms, were the first in the world to organise their own professional R & D laboratories in the 1870s (Haber, 1958). The growth of organised research and development by industry was a major innovation which spread to the US in the 1890s and to most industries in other advanced countries during the first half of this century. During the industrial revolution, R & D was the business of individual inventors and innovation of entrepreneurs, in some cases (for example Perkin, Baekeland, Nobel, Solvay and Linde) the same individuals. In the German chemical industry in the 1870s, and gradually elsewhere, both activities became largely the business of established firms.

During the First World War the UK government became keen to foster an alliance between the state and science, in its search for improvements to military technology, and in its need to find the means to fight a war against the country which supplied most of the world's optical equipment, magnetos, dyes, drugs and acetone (for explosives). With the championship of Lord Haldane (First Secretary of State for War and later Lord Chancellor) the relationship between science and government for the first time developed beyond an *ad hoc* wartime arrangement (Poole and Andrews, 1972). Science advisory committees and the 'beginnings of a sketchily integrated science policy' (Rose and Rose, 1970) were established in 1914.

By 1916 the pressure for an independent ministry of industry and for a government department responsible for directing funds towards research and development or establishing institutions to study various technological and industrial problems, led to the acceptance of Haldane's proposal for a Department of Scientific and Industrial Research (DSIR). The DSIR was responsible to Parliament, had its own funds voted by parliament, and was the major source of state support for science and technology in Britain for nearly fifty years. This represented the beginnings of *direct* government involvement in technology.

The DSIR provided research studentships for higher degrees and promoted technological change in industry by encouraging the establishment of cooperative research associations by industry. For every £1 invested by industry in the research associations, another £1 was provided by the government. During the 1920s and 1930s the DSIR expanded its support for these activities and began to develop totally government funded research, for example in the National Physical Laboratory, Road Research Laboratory, Low Temperature Research station and other research establishments. In 1939 its annual budget was about three times the 1919 figure, but still only 0.1 per cent of GNP (compared to 0.6 per cent in the US) (Rose and Rose, 1970).

In 1917 a group of university, government and industrial scientists established a professional association for scientific workers, the National Union of Scientific workers (NUSW), which later became the Association of Scientific Workers (AScW)(and is now ASTMS). During the 1920s and 1930s this organisation campaigned for the improvement of scientists' status and pay, and for more positive direction of science towards social advance.

J. D. Bernal, chairman of the AScW executive, published *The Social Function of Science* in 1939. He was the most well-known of many advocates of economic planning, and the planning of science and technology, during this period. The idea of planning or promoting technology by government was, however, strongly associated with the radical political views of Bernal and his associates in the AScW such as J. B. S. Haldane (nephew of Lord Haldane), Hyman Levy, Lancelot Hogben, Joseph Needham and Julian Huxley and received little echo (and quite a lot of hostility) in government, or even wider scientific circles, until the Second World War.

Opponents included Karl Popper, Michael Polanyi who formed the opposing 'Society for Freedom in Science' and John Jewkes, who later wrote *Ordeal by Planning* (1948) before contributing to *The Sources of Invention* (Jewkes, Sawers and Stillerman 1958). The debates of this period are more thoroughly discussed by Werskey (1978).

But, opposition to planning science was swept away by the Second World War. Government-sponsored war research developed very rapidly, although many aspects remained hidden from parliamentary scrutiny, such as the decision to develop atomic weapons. Huge increases of investment in R & D led to the expansion of centralised bodies for scientific decision-making. Scientists were mobilised for the war effort. The notion of permanent governmental policy-making bodies for R & D after the war gained support. The development of radar, penicillin and the atomic bomb and the establishment of operations research are all examples of successful 'big science' projects in the allied war effort. These successes contributed greatly to the respectability of planning and science policy, no longer seen as a left-wing proposition half-way to communism. It was also these successes which contributed to the previously discussed hegemony for a number of years of 'science-push' ideas in science and technology policy.

By 1950 the governments of most advanced countries had adopted policies aimed at deliberately fostering economic growth. For example, the Beveridge Report had been based on Keynesian assumptions that the National Health Service and Welfare Services would stimulate the economy. Above all, it was believed that growth was the result of a high volume of investment, particularly in innovation. Consequently policies were, in most cases for the first time, adopted for fostering science and technology in order to increase the volume and variety of output and the productivity of labour. Science and technology began to account for an increasing percentage of GNP and to employ an increasing number of people. It is probably true to say, however, that the link between technology and growth was implicitly assumed rather than explicitly stated. No sophisticated analysis by governments had yet explored the relationships and few economists treated technical change as anything but exogenous. It seemed that higher expenditure on science and technology was accepted in the 1950s and early 1960s as an unfailing recipe for technical progress and increased wealth: it seemed that 'science-push' worked.

But it should not be thought that developments at the level of the nation-state, defence policy, and scientific lobbying were the only influences. By 1950 the firm, too, was changing. Large multiproduct, multi-divisional, multi-national enterprises with turnovers of the same order of magnitude as small advanced countries (for example Belgium or Holland) were proliferating and exercising increasing influence over the economies of nation states. The increasing complexity of their influence on technology has been discussed in Parts I and II. If governments wanted to direct technology, it was clear they would need a variety of complex policy instruments aimed at directing these corporations. And if the corporations did not want to be 'directed', the governments might not be able to make them.

We have spent some time analysing the circumstances surrounding the first attempts at direct government intervention in technological change, and how this experience laid the basis for the way in which government direction of technology developed afterwards. These approaches set the tone for the post-Second World War attempts at promotion and control of technology.

The nature of markets, the firm and the state, and the relationship between them have dramatically changed in the last one hundred and fifty years. The direction of technology, by each of these institutions in various and changing ways, has evolved accordingly.

Since the early campaigns for particular government intervention (such as the Alkali Acts), the arguments for and against direct government intervention in general, and about the extent of that intervention, have become much more structured. It is also noticeable that attitudes, particularly those of industry and government, have tended to change with structural economic changes. Social reforms, including regulations of the unwanted consequences of technology, have been concentrated in the periods of economic upswing after 1870 and after 1945. In periods of economic recession, when the material basis for concessions has been eroded, social legislation and welfare state ideas have been more widely attacked. The traditional government balancing act between 'efficiency' and 'equity' (as they would see it) will be tipped toward considerations of 'efficiency' and a greater emphasis on competition and on promoting the institutional changes that permit structural change. In some cases, promotion of technology may suffer as much as control, as has arguably been the case in the UK in the early 1980s.

Chapter 9 will develop some of the points raised in this chapter and will give a more detailed account of government policies towards science and technology in the post-World War Two period.

9 The Promotion and Control of Technology by Government

9.1 INTRODUCTION

We have seen that government involvement in technical change can be usefully divided into policies for promotion and for control of technology (Johnston and Gummett, 1979), and that these policies may interact in various ways. It is also clear from the historical development of these policies that they owe a great deal to the prevailing thrust of government policy with respect to the economy and the social order, as well as to specific concerns with particular aspects of science and technology. Thus promotion policies, for example, can be divided into direct policies which target specific institutions responsible for technical change, and indirect policies which have a broader focus on the business environment.

With the rapid growth in funding of science and technology after the Second World War, and the growth of all economic and industrial policy, largely as a result of the enthusiasm for Keynesian demand management, government involvement in technical change expanded considerably. This chapter concentrates on this post-war period. The first part of the chapter presents an account of the promotion policies of UK Labour and Conservative administrations, and uses some case studies to illustrate similarities and differences between the approaches adopted. The second part of the chapter discusses some of the issues involved in the control of technology with examples from post-war developments. The powerful effects of institutions outside the market and the firm are the common theme of these analyses.

9.2   LABOUR GOVERNMENT POLICIES FOR PROMOTION OF TECHNOLOGY

The 'science-push' philosophy discussed in Chapter 8 reached the height of its influence at the time of the 1964 general election in the UK. Consequently the role of government in ensuring economic growth by promoting science and technology was a more central issue in that election than in any election before – or since. Both major parties recognised the importance of technology but the Labour Party in particular had a technocratic dimension to its propaganda. One of Harold Wilson's most well known remarks was his reference to 'the Britain that is going to be forged in the white heat of this technological revolution' (Labour Party, 1963).

In retrospect it may be seen that the Government's understanding of how it was going to forge a new Britain in the white heat of a technological revolution was not very precise. In the event its technology policy evolved, rather than being established from the beginning. The Department of Economic Affairs was created in 1964 to oversee the National Plan, but this effectively withered away after the 1966 economic crisis. The Ministry of Technology, known as Mintech, was also set up in 1964 as the main agency for promotion of technology and in contrast with the Department of Economic Affairs it grew and grew and effectively evolved into a ministry of industry. By 1969 it had nearly 40 000 staff compared to a figure of under 6000 in 1964, as a result of mergers with other departments (such as the Ministry of Aviation and the Ministry of Power) and the assumption of responsibilities previously the province of other departments (such as the control of the Industrial Reorganisation Corporation or the sponsorship of various industrial sectors).

But while Mintech grew, so did the paradoxical discrepancy between Britain's growth rate which was not as great as that of Japan, France, Italy or the German Federal Republic, and its expenditure on R & D, which was higher than that of any other European country. This raised the suspicion that perhaps the encouragement of more R & D was not by itself necessarily an unfailing recipe for economic success. During this period the study of science policy and innovation began to be accepted as a valid academic discipline with the establishment of university departments or research units in the area. The government was prepared to fund research which promised to uncover information about the nature of the innovation process and the relationship between R & D, innovation and growth. It was aware, with the slowing down of economic growth, that it needed to cut back on R & D expenditure or be convinced that a continued increase really would lead to an increase in growth. Meanwhile, industry was concerned about its own expenditure in R & D, the relationship between that and profitability, and about more deliberate and cost-effective ways of managing the R & D process. Concern with innovation and R & D developed in the management literature with the appearance of prescriptions for improved project selection and evaluation and the management of innovation in general (for example, Ansoff and Stewart, 1967). In the event, R & D expenditure did fall, but perhaps for other reasons. Policies of *détente* softened the cold war, and the urgency of military R & D was not seen as being so great. Military expenditure (as a percentage of total R & D) fell in most countries during the 1960s; and after the moon landing, space R & D in the US was also cut back. At the same time the scale of government-funded R & D, some well publicised cases of vast cost over-runs on big science projects, and the beginnings of concern about some of the side effects of technological change (such as pollution), were giving rise to demands for more public account-ability, while the 'spin-off' justification – that military R & D had a beneficial effect on the economy and technology in general – was beginning to be discredited.

Williams *et al.* (1982) sum up the 1964–70 Labour Government's attempts at promotion of technology with the following reference to the Ministry of Technology responsible for the job:

> The rapidity of Mintech's growth, the expectations and pres-sures which derived from the commitments it inherited, and the fact that everyone was really 'feeling in the dark' so far as a technology policy was concerned, all contributed to an approach which was always energetic; this did, however, exhibit only limited relevance and sometimes seems to have been derived from conflicting principles.

As the rate of R & D growth began to slow down, more concern was expressed about the results of R & D rather than just the amount of it. Big science projects were more closely examined to try and avoid wastefulness, though most of these projects still went ahead. The case study research on innovation (for example Langrish *et al.* 1972: SPRU, 1972) was beginning to indicate that the relationship between R & D, discovery, innovation and growth was not simple or linear; however the emphasis in these studies on customer needs and communication with the market was interpreted by policy makers as a turn towards 'demand-pull' theories of innovation (see Chapter 5).

Thus, between 1964 and 1970 the Labour Government erected a substantial apparatus to pursue its original 'science-push' philosophy, yet by the end of the period, the philosophy was being upturned, and the tide of R & D expansion seemed also to be on the turn with a consequent slowdown in the growth rate of R & D expenditure.

Labour was out of office between 1970 and 1974. When it returned to power the issues of promotion of technology were much more intimately influenced by those of control because public concern over the potentially negative effects of technology had increased considerably. This concern was exemplified in the debates on pollution, on working conditions, on the costs of growth and other topics which are taken up later in this chapter and in Chapter 10. Furthermore, Mintech and its responsibilities had been absorbed into the Ministry of Industry, reflecting the view that technology policy and industry policy were too interdependent to be managed by two departments. Despite the new environment, the growth objective still predominated. The proportion of UK government R & D funds spent on defence and aerospace grew again throughout the 1970s, while in other advanced countries defence and aerospace R & D declined proportionately. However some moves in the direction of increased public accountability were attempted.

In a new initiative taken under the Industry Act (1975) the government set up the National Enterprise Board (NEB) 'to promote industrial efficiency and competitiveness' (Department of Industry, 1975a). A further provision of the Act was the idea of planning agreements, which required firms to make planning information available to the government and in some cases to trade unions, so that corporate policy might be influenced by the interests of groups of people – such as employees or customers – who, in addition to the shareholders, would be affected by the activities and plans of the firms concerned. This policy, which was more strongly supported on the left of the Labour Party than the right, can perhaps be seen as an attempt explicitly to combine promotion and control. The policy made little headway however, meeting resistance from industry, and eventually being dropped following a cabinet reshuffle which removed the Minister providing its main motivation.

A more pragmatic and developed aspect of Labour Government policy following the 1975 act was that of 'picking winners': identifying companies which could become successful with government finance (or those whose failure would have repercussions too serious to be allowed). The National Economic Development Council established a number of sector working parties (covering 39 sectors by 1977) whose function was to analyse their particular industrial sector and make detailed policy proposals. This was the high point of Labour Party *dirigisme* in the 1970s.

Meanwhile, researchers were beginning to elaborate more sophisticated theories about the relationship between technological and economic change, and between innovation and Britain's competitive position. Many studies demonstrated that R & D is one of many inputs to the innovation process, and that successful innovation depends on many things besides good R & D, including management skills and an understanding of the market (Chapter 5).

Pavitt and Soete (1980) showed a statistical correlation between countries' levels of innovative activities and their world export shares, in most categories of capital goods and chemicals and more weakly in durable consumer goods and iron and steel. In several industrial sectors it was shown that innovation led to growing market shares, while lack of innovation led to decline.

The lack of correlation between R & D expenditure and economic growth, noticed earlier, was partly the result of high levels of spending on defence R & D in Britain. Kaldor (1980) argued that the high level of innovative activity in the defence sector was an integral part of the relative decline in the capacity for civilian innovation in British industry. She argued that it was both a cause and a consequence of the decline; a consequence in that defence markets often secure outlets for the production of uncompetitive firms; and also a cause, since resources and energies are concentrated on unnecessary technological embellishments, while the neglect of users' needs and market constraints are allowed to persist.

In the civil field, it was found that Britain was producing relatively unsophisticated machinery and durable consumer goods, requiring relatively few innovative activities and having relatively low unit values and value to weight ratios, compared to Germany and other major competitors (see Section 6.8 above). Pavitt (1980) found that firms and countries gained a competitive advantage when they deliberately and successfully committed more resources to industrial innovation, but that innovation did not follow automatically from the 'right' volume of investment. The relative lack of technical and professional competence of management in British industry was also seen as significant (Allen, 1976), and a problem which could be dated back to the First World War (Walker, 1980).

In summary, these research results (which rested on a broader base, as reported in previous chapters) led innovation and technology policy researchers to make a number of recommendations. They proposed a reduction to more modest proportions of R & D programmes in the defence, aerospace and nuclear sectors, and much greater investment in the skills and innovative activities of 'bread and butter' sectors. Industry was urged to pay far more attention to market intelligence and to good communications between production engineering, product design, marketing and management. The need for more and better professionally and technologically competent managers and engineers in manufacturing industry was echoed by a Government Report (Committee of Inquiry into the Engineering Profession 1980). Government was urged to supplement the activities of industrial firms in improving the technical quality and international competitiveness of manufactured goods.

Whilst these recommendations to government from the research community were not by any means the most powerful influence on policy, they did broadly coincide with the development of government opinion. Simple science-push and demand-pull approaches to the role of government were beginning to be replaced by a more subtle view in which the intention was to combine specific stimuli for technology with broad policies aimed at creating favourable environments for innovation and more effective management. But translating intentions into policy instruments remained difficult and controversial. Whatever the policies of post-war Labour governments much of the practice has been formulated on an *ad hoc* basis in response to the pressure of events and often at variance with the stated aims. We turn next to the policies of Conservative administrations during the post-war period, and then to some examples of policy problems which have exercised governments of both parties.

9.3 CONSERVATIVE POLICIES FOR PROMOTION OF TECHNOLOGY

The common features of the 1970–74 and 1979–present Conservative Government technology policy have been 'disengagement' and *laissez-innover*, practiced with more determination by the Thatcher than by the Heath Government. The Heath Government proposed a series of measures for reducing the size of the public sector and sharpening competitive pressures. Thus, for example, the Monopolies Commission was given greater powers and attempted to weaken the monopolies of firms such as Hoffman La Roche (makers of valium) who dominated the tranquilliser market. It preferred to intervene indirectly rather than directly, by influencing industry's environment. Despite this overall aim, however, the government recognised the limitations of market forces in affecting investment in technologically-based industry – even market forces orchestrated by government policies. Even at the policy proposal stage before the 1970 election the qualification was made that some industries (aircraft, shipbuilding and electronics) would require special treatment, as the cost of high technology was so great that the market could not be expected to supply sufficient funds (Joseph, 1970). But despite these qualifications disengagement was real enough in some sectors. A number of firms were refused government aid and were allowed to go bankrupt. Labour's plans for nationalisation of some sectors were cancelled. Private capital was introduced into some public industries while subsidiary activities of other public industries were sold off to private buyers.

However, the government was finally obliged to reverse this policy in relation to Rolls Royce. Rather than allow the firm to go bankrupt the Conservative government not only made large sums of public money available, but actually nationalised it. The Secretary of State for Trade and Industry, John Davies, made it clear that government intervention was justifiable where private investors were not prepared to take the risks associated with the necessary investment, or wait for long term returns, or where an industry was in a key position regarding international competition or employment. The 1972 Industry Act allowed new and selective support for industry to compensate for private capital's reluctance to invest under these circumstances. Thus in practice the Conservative government ended up with a policy very similar to that of the previous Labour government.

After the 1974–79 Labour government the Conservatives returned under Margaret Thatcher, as critical of the policy reverse under Heath as they were of Labour. The Thatcher government's policy towards technology and industry has represented a more radical rejection of the consensus Keynesian approaches followed (to a greater or lesser extent) by other post-war governments of both parties. Macroeconomic policies have attempted to control the money supply and reduce the public sector borrowing requirement in order to reduce inflation. 'Competition' and 'the free market economy' have been the key words, with privatisation introduced into several sectors and public monopolies ended. For example, the most profitable sector of the Post Office, the telecommunications business, was first made into a separate Public Corporation (British Telecom); it then lost its monopoly in supply of telephone and telecommunications equipment; and more recently it has been privatised.

The National Enterprise Board was seen as a symbol of the previous Labour government, and the Conservatives promised before the 1979 election to wind it up. For a time, however, Industry Secretary Joseph allowed it to continue with limited powers. Though Joseph is an ardent supporter of *laissez-faire* in principle, he appeared to recognise the limits of market forces. Firms in difficulty might have to be supported by the NEB, he said, if they were viable in the long run, but could not find a solution in the private sector.

Eventually the NEB was merged with the National Research Development Corporation (NRDC) to form the British Technology Group. Since September 1983, BTG has been selling off the former NEB's shares in some 25 companies, making a profit of £34m by August 1985 (*New Scientist*, 1985a). At the same time, loans of up to £50 000 were being offered to stimulate new businesses, especially in the area of high technology. The BTG's role is thus seen by the government as a vehicle for the reduction of direct, and increase in indirect involvement in technology and industry.

For about thirty years the NRDC had a monopoly on the exploitation of inventions or discoveries made as a result of government-funded research. Profits from such inventions amounted to £25m in 1983, £14m in 1984, and £18m in 1985 (New Scientist, 1985a) the most successful being the cephalosporin drugs (patent expired 1983) and pyrethrin insecticides (new patent 1984). This monopoly has now been ended in an effort to create a more fluid interaction between public-sector science and the private sector.

But the attempts to promote the entrepreneurial application of science have been combined with severe constraints on basic research. The Medical Research Council, for example, is now only able to fund 53 per cent of what it considers to be alpha-rated research (Connor, 1984).

Overall however, attempts at disengagement could not be said to be much more successful in the 1980s than they were in the early 1970s. Certainly severe cuts in spending on R & D, and privatisation of industry and services have taken place: but public spending has not conspicuously diminished. It has increased significantly in the military and prestige 'big technology' sectors once again. In 1984 the share of government R & D funds going on defence projects was higher than ever. The Reagan administration in the US has similarly increased the Pentagon's share of R & D funds, which went up by 27 per cent in 1985 (Joyce, 1985). Of the Science and Engineering Research Council funds, 'Big Science' consumes almost double the resources of all the remaining natural sciences together (Martin and Irvine, 1984). British spending on R & D is still one of the highest in the world as a proportion of GDP – but since Britain's GDP has declined relative to its major competitors this represents a decline in R & D spending; and according to Martin, Irvine and Turner (1984) a qualitative decline in results. The increase in military R & D also means a greater relative decline in government spending on civilian projects. Britain now spends a higher proportion of its GDP on military R & D than any other Western nation (*New Scientist*, 1985b) while having the dubious distinction of being the only Western nation with an absolute decline in R & D performed in manufacturing industry, paid for by industry or government (OECD 1985).

But effects should not be confused with intentions, and it is clear that the Thatcher government has not been unconcerned about innovation. On the contrary it has expressed its belief in the importance of innovation in economic growth, competitiveness and the creation of wealth. In September 1983 the Prime Minister organised a seminar on science, technology and industry. The government's view, however, is that innovation is the business of entrepreneurs and the government's role is to provide the sort of economic environment which will stimulate market forces in such a way that the innovation process will be encouraged. The potential for de-regulation and freer markets to stimulate technical change in *some* sectors should not be underestimated, although its desirability remains an ideological issue. But, as much previous policy experience suggests, it must be considered untenable as a general solution to the needs of technical change in *all* sectors, quite apart from ideological considerations. However, this problem is not high on the agenda of government in the 1980s.

In summary, the Conservative governments of 1970–74 and 1979 to the present day have attempted to stimulate innovation and technical change by policies designed to increase competition, revitalise market forces and decrease direct intervention. The present government has pursued a 'disengagement' policy more successfully than the Heath Government, in that it has refused to use public money to help several companies in difficulties and has succeeded in privatising some areas previously publicly owned. However, in an era of multinational companies, slow economic growth, big spending on R & D (compared to earlier historic periods) and the reluctance of private firms to take the risks of investment in some types of innovation, the government is still heavily committed to public R & D spending, and though it has been reduced in the civil sector it has certainly been balanced by an increase in the military sector.

Thus many of the institutions of direct and indirect government promotion of technology must be presumed to be so firmly embedded in the structure of UK society that they are not easily dismantled. The policy stance of Conservative governments nevertheless substantially alters the effects of these institutions. It is therefore not a question of whether to intervene, but how?

9.4   THE EXAMPLES OF AERO ENGINES AND MOTOR VEHICLES

The aero engine and motor vehicle industries provide an interesting contrast in illustrating the approaches of Conservative and Labour administrations towards technologically-based industry. They also provide a contrast in the very different approaches of government, independent of party politics, to these two industries. The manufacture of aero engines was recognised as strategically significant for the UK by Labour and Conservative governments, and policies adopted accordingly.

Rolls Royce was first backed by the Labour Government when in 1968 it started developing the new RB211 engine to be supplied to Lockheed (Williams *et al.*, 1982). When the project subsequently ran into financial difficulties the Labour Government continued to assist Rolls Royce and there was no substantial change of policy, certainly not in the sense of a greater disengagement, when a Conservative government was elected in 1970. Indeed, it was under the new Conservative Government that Rolls Royce was nationalised in 1971. That no substantial difference in the approach to the aero engine industry existed between the Labour and Conservative governments is probably partly due to the nature of the aero engine industry. Both parties had in fact acknowledged the central importance of the industry for prestige and strategic reasons. A policy of complete disengagement was therefore unthinkable. The only difference between the Labour and Conservative parties could be the mechanisms of intervention chosen. In addition, the habit created by the previous support and the continuing power of a lobby determined a preferential treatment of R & D in the aerospace sector which went far beyond a considered assessment of transport needs (Freeman, 1982).

The motor industry is also a sector of major economic significance in Britain and elsewhere, for reasons detailed by the Central Policy Review Staff (CPRS, 1975) and by Stubbs (1979). During the 1960s, the British motor industry began to decline, investment in R & D and machinery were lower than that of foreign competitors, yet the Labour Government did not intervene, despite its overall philosophy described in the previous section. This was partly due to the fact that governments had not previously supported the industry, unlike the aerospace and defence sectors where there was already a precedent for large-scale government spending. In addition, Labour's technology policy began at a less interesting stage of the motor industry's life cycle – characterised by incremental and process innovations rather than radical product innovations, and intervention was less easy to justify since the government was not the primary customer (Stubbs, 1979).

When Labour was re-elected in 1974 the decline in the motor industry was much more serious than it had been in 1964–70. When in power previously, Labour had allowed the takeover of Rootes by the US Chrysler Corporation in 1967 and encouraged the BMC–Leyland merger in 1968. In 1975 the Labour Government saved British Leyland from bankruptcy by taking over the company and giving it to the National Enterprise Board with £1000m to spend on capital investment (Department of Industry, 1975b). The interests of UK component manufacturers were an important consideration here.

The example of British Leyland, although it ended in the nationalisation of the firm, shows considerable differences in government attitudes toward the motor industry with respect to the aerospace industry. An even more striking difference is shown by the Labour Government's rescue of Chrysler in 1975. The Labour Government, though generally committed to state intervention and (to a limited extent) to public ownership, rejected the option of nationalisation. Instead they negotiated a rescue package with Chrysler Corporation (the US parent) without even a government equity holding, simply in exchange for a formal undertaking to remain in Britain. The Labour Government justified the rescue on the grounds of employment, of the credibility of British firms in overseas markets, and of the possible threat represented by the Scottish National Party, given that most jobs at stake were in Scotland (Williams *et al.*, 1982). The importance of these factors cannot be denied, but it is doubtful whether they were real causes of the policies adopted or whether they were used to justify policies already chosen (although implicitly) on other grounds. It is difficult to believe that the same hesitation and lack of commitment that were shown toward the motor industry would have been present if the firms at stake had been in the aerospace sector. From this analysis at least three factors emerge as important determinants of industrial and technology policy: the political philosophies of different governments, the nature of the industry, and the policies previously adopted. The interplay of these three factors can sometimes lead to a policy convergence between governments of different political persuasion.

A fourth factor, the multinational nature of the firms involved, differentiated the case of Chrysler from those of Rolls Royce and British Leyland, which were essentially British-based firms. The bargaining power of the UK government was likely to be reduced when dealing with a foreign-owned multinational used to worldwide sourcing.

Labour's experience at intervention in the electronics industry (next section) did not involve attempts to persuade a foreign-owned multinational to cooperate. It was an attempt to bridge a gap in UK technological capability by investing in a new, small firm, and provides the next example of technology policy in practice.

9.5 GOVERNMENT POLICY AND MICROELECTRONICS

Almost every industrial and service activity has the potential to be influenced to some extent by the 'microelectronic revolution', although the actual rate of diffusion of the new technology depends on various enabling educational, social and managerial changes (Freeman, 1984 and Chapter 7, this volume).

When the NEB was established by the Labour Government in 1975, its general goal was to develop the UK economy and increase its competitiveness. It was to concentrate its resources on a few key sectors of the economy, of which microelectronics was a clear candidate. At its formation the NEB acquired existing government shareholdings in a number of firms in the electronics sector, including ICL, Ferranti and Data Recording Instruments. With an interest in mainframe computers, small computers, electronic systems, integrated circuit manufacture and computer peripherals (Willott, 1981) it seemed reasonable for the NEB to develop a coherent strategy for the whole sector. Meanwhile, the NEDC electronic components sector working party, as part of the policy of 'picking winners' had identified the sector as one with growth potential.

The convergence of communications and computers was bringing computer-based products into general business, and later domestic, applications and away from just specialist uses. With rapidly falling prices and more and more functions being performed by smaller and smaller devices the potential market seemed enormous. The dominance of US giants like IBM had been slightly reduced in the mid-1970s by their slowness to move into these new markets (Willott, 1981), thus presenting an opportunity for British firms. Yet at this stage, UK government support for the electronics sector was far less than that of France, Germany the US and Japan. Thus a CPRS (1978) report recommended substantial government support to enable British firms to innovate in order to reach a competitive position in the market. In July 1978 a £70m programme of government assistance in the development and manufacture of microelectronics products was announced. The NEB was to invest £50m in the silicon chip firm Inmos (Williams *et al.*, 1982). This was a 'green field' venture to get in at the changeover from large scale integrated circuits (LSI) to very large scale integrated circuits (VLSI).

Until the late 1970s the electronic components market was dominated by the US multinationals which had diversified into civilian markets after their rapid growth in the 1960s on the basis of military and aerospace contracts. These firms had 90 per cent of the UK market in 1978. Since then, Japanese firms (with $350m government support for R & D) have caught up with the US firms. Of British firms, only Ferranti, Plessey, GEC and Lucas were making chips in 1978, all concentrating on 'custom made' or 'special' chips. UK users of 'standard' chips bought them on the world market, but were left at a disadvantage when there was a world shortage. They were also unable to benefit from close contact with the manufacturers' R & D activities. There was thus strong support for the idea of a UK manufacturer of standard chips, and for the government to back such a venture. Inmos was set up in July 1978, having received its first £25m from the NEB, the other £25m to be paid in 1980 if the firm was on target. The NEB was thus prepared to take risks and 'back a radically different course from the received wisdom of the industry' (Willott, 1981).

In 1979, the Conservatives were elected to power and Keith Joseph became Industry Secretary. Although microelectronics was just the sort of industry that Joseph thought Britain should be developing, he was opposed to the expenditure of public money to do so. It was his belief that private sources of capital should be used for such a scheme. The problem was that private capital was not forth-coming.

In November 1979 Inmos applied for its second £25m, saying it was on target, and due to create 3000 jobs in Britain by 1984 and contribute £95m to net exports (Williams, *et al.*, 1982). Eventually in July 1980 the NEB was allowed to invest the second £25m in Inmos, (which had been agreed in principle the previous July) on condition that the factory was built in South Wales. Thus, Inmos became the only general purpose silicon chip producer in Britain.

However, the government continued to believe that this was an activity for private firms, rather than the government to invest in, and decided to sell its Inmos shareholding. From 1980 the NEB began trying to sell off its assets. It sold its 25 per cent interest in ICL for £38m, but only fifteen months after the sale the government found itself supporting ICL, then in serious financial difficulties, with a loan guarantee of £200m (Williams, 1982). The NEB was forced to increase its funding to Rolls Royce and BL, and generally found itself financing nationalised industries at a higher level than was consistent with the government's ideological viewpoint. One of the main difficulties was finding private sources of capital, and selling Inmos took longer than expected. At one time it was feared that the only willing buyers would be foreign-owned companies. Finally Thorn EMI paid £95m for Inmos in September 1984, BTG realising £30m profit on the sale (*Dulyell*, 1985).

At the beginning of 1985 Inmos ran into technical difficulties with the production of Dynamic Random Access Memory (DRAM) chips. It has now stopped DRAM production and Thorn has set aside £27m to cover Inmos' losses plus the cost of 200 redundancies. At the time of writing, the story of Inmos as one example of government attempts to promote innovation in micro-electronic technology is still continuing. It could be argued from the Labour viewpoint that the sponsorship of Inmos was 'too little, too late'. The Conservative decision to re-sell its interests in Inmos was certainly consistent with its general philosophy, however.

Both policies aimed to increase the competitiveness of the *country* in microelectronics, but encountered the problem that private capital investment in the sector is influenced by a wider variety of considerations.

9.6   SOME PRE-CONDITIONS FOR GOVERNMENT CONTROL OF TECHNOLOGY

The following sections analyse government attempts to control the effects of technological change in terms of the mechanisms for regulation: criminal law, civil law and voluntary agreement. Other structures exist to ensure that these laws or agreements are implemented, in the form of inspectors, licensing authorities or bodies giving planning permission. Yet other bodies are established to give early warning of the need for regulation, assess the possible future consequences of technical change and advise legislators. We also consider the assumptions made in the assessment of risk and classify the types of unwanted effects in terms of potential hazards: for example, product safety, health and safety at work, or hazards of the general environment. First of all, however, it is worth pointing out that attempts to control these unwanted effects arose in a very piecemeal and haphazard way, as the problems themselves were recognised, and in many cases success was strongly dependent on technological changes which themselves permitted the recognition of the problem. Consideration of some of these pre-conditions for effective regulation will serve as an introduction to the analysis of government regulation.

Historically, legislation has taken place as a response to the recognition of an actual (not potential) hazard or other problem; in many cases the recognition itself, or the identification of the cause, has taken many years. The protracted introduction of regulations governing the use of asbestos is a good example of this. Asbestos was first mined and processed before 1900, but the recognition of asbestosis as an industrial disease caused by exposure to the material did not happen until 1930, when the Merewether Report was produced for the Home Office (Merewether and Price, 1934). Regulations were introduced in 1931. Recognition of lung cancer as an industrial disease caused by exposure to asbestos did not take place until 1955 (Doll, 1955), while mesothelioma, a rare pleural cancer, was not established as an asbestos-related disease until the mid-1960s. The 1931 regulations were not amended until 1969. Further amendments to the regulations were still being made as recently as 1984 (Health and Safety Executive, 1984).

Another pre-condition for successful regulation is the existence of technology for detecting the presence of the hazardous material. Concentration of toxic materials in the environment cannot be regulated if the materials cannot be detected or the amounts present measured. The post-World War II developments in spectroscopy and chromatography were thus of major importance in making detection, measurement and therefore regulation of toxic materials possible. However, regulations requiring zero concentration of toxic materials become technically and economically impossible to implement once microscopic quantities of materials can be detected. Hence specific limits have to be set, usually in parts per million.

Another scientific advance which permitted more thorough regulation of an unsafe environment, by establishing causes of ill-health, was the development of epidemiology. It is very difficult to establish a causal relationship between exposure to a given material and contraction of a disease in individual cases. For example, mesothelioma can take thirty years to develop after a very short exposure to a small quantity of asbestos. Many diseases may be caused by a variety of factors. Epidemiology establishes a statistical relationship between various events and thus a probability that a given event caused a particular illness. Recently, for example, attempts have been made to establish links between cancer and environmental factors by the plotting of 'cancer maps', showing areas of high evidence of cancer (MacKenzie, 1984).

The development of the science of ecology provides another example of a scientific or technological advance that permitted more effective regulation. Ecology is based on the inter-relationships between all parts of the natural environment, and its development has tended to cut across the fragmentation of science into specialisms. For example, at one time it was usual to study the effects of a particular chemical on an insect, without observing the consequences of those effects for plant life, other animals and human beings. Such an approach made possible the widespread use of pesticides without understanding their effect in polluting the environment. Today, an ecological approach taking into account the possibility of unwanted effects is an essential component of pesticide research.

Sometimes a pre-condition for governmental action to control the unwanted effects of technological change has been a major disaster. For example, the thalidomide disaster in the 1950s was a major factor in getting the Kefauver–Harris Amendment to the US Food, Drug and Cosmetic Act passed, and the Dunlop Committee on Drug Safety established in Britain. The UK Advisory Committee on Major Hazards was set up by the Health and Safety Commission after the Flixborough explosion in 1974. Arguments in favour of stricter drug licensing or guidelines for control of potentially dangerous plant had been made long before, but it took a disaster to spur the government to action.

9.7 CONTROL OF TECHNOLOGY: ATTEMPTS TO ANTICIPATE THE NEED FOR REGULATIONS

The problem with control triggered by reaction to events is that, by the time the undesirable consequences of a technology are discovered, the technology is often so much a part of the whole economic and social fabric that its control is extremely difficult (Collingridge, 1980). However, some attempts have been made to identify potential hazards in advance, before damage is done, and while the technology is neither too highly developed or diffused to be easily changed. Technology assessment is intended to be just such an activity. It aims to identify all the possible impacts of a new technology and not just the intended ones; and it tries to include effects that cannot be measured quantitatively as well as those that can. (Hetman, 1973).

The Office of Technology Assessment (OTA) was established in the US in 1970 and by the mid-1970s had a budget of $8 million and employed one hundred professionals. The OTA was set up after Congress voted against the supersonic transport Lockheed L2000, intended to rival Concorde. Congress wanted independent advice and thought that the expert witnesses representing the Nixon administration and the aerospace industry made out a very biased case. The OTA was to aid Congress in making decisions on technological matters (Roy, 1982).

Under the US National Environmental Protection Act (1970) 'environmental impact statements' are required to indicate what impact any federally-funded project will have on the natural and socio-economic environment. The OTA has recently reported on acid rain, the commercial implications of biotechnology, and the strategic defense initiative ('star wars'). OTA reports identify areas of doubt and uncertainty and whether something can or cannot be proved scientifically; and outline options for legislation, indicating what can be done and the cost of doing it (Lloyd, 1985).

Several European Governments are currently following the example of the US in setting up offices of technology assessment, but Britain is not one of them. However, although technology assessment is not formally incorporated in planning and policy making in Britain, attempts at technology assessment, by other names, are nevertheless made in practice. The 'Windscale Enquiry', concerned with a nuclear fuel re-processing plant, is an example of this.

9.8   MECHANISMS FOR THE CONTROL OF TECHNOLOGY

There is a spectrum of mechanisms for the control of technology, ranging from reliance on market forces (no government or local authority intervention at all) to complete removal of decisions on control from the private to the public domain (nationalisation of the firms or industries responsible for the negative effects of innovation) and hence removal of at least some of the conflict between social costs and private benefits.

In between, and more commonly, regulations of specific features of products or processes may be achieved by criminal law, civil law or agreements between government authorities, and the establishment responsible for the technical change. Prosecution under Criminal Law has the advantage that the forces of the state are employed to protect society from the unwanted effects of technical change; but it has the disadvantage that legislation takes time to enact, is not very flexible to changes in acceptability of risk or measurement of hazards, and is dependent upon the (costly) existence of some sort of organisation established to monitor and implement the legislation. Thus, for example, the UK Health and Safety Executive are responsible for implementing pollution and health and safety legislation, while the UK Department of Health and Social Security Medicines Division, advised by the Committee on Safety of Medicines, are responsible for establishing the safety of new drugs and granting licenses for their clinical use.

The legislation is only as effective as the agencies responsible for implementing it. The number of Health and Safety Executive (HSE) staff fell from 4500 to 3800 between 1980 and 1983 even though the Chairman of the Health and Safety Commission informed the government in May 1980 that 'any further reductions' in staff would make it impossible for the HSE 'to carry out its full range of functions'. (BSSRS, 1983).

Civil law is perhaps more flexible, but in other ways less satisfactory. It is the responsibility of the individual who has been harmed in some way to sue the person or organisation responsible for damages. This implies having the money (or legal aid) necessary to take legal action and often means proving that you have been harmed and proving negligence. In the ten years of litigation surrounding the thalidomide case it was necessary to prove legally (not exactly the same as proving scientifically) that thalidomide was responsible for causing foetal deformation, and that the manufacturers were responsible, since it was not a legal requirement at the time to test drugs for teratogenicity. Voluntary Agreements (for example between Health and Safety Executive and employers on safety standards and safe practices for a particular industry) are even more flexible as they can be amended whenever new evidence about the safety of particular materials or work practices become available. However, where there is disagreement or conflict between different interests involved, such agreements are not legally enforceable.

It is also possible to regulate unwanted side effects indirectly, using financial incentives such as an effluent tax levied on manufacturers or an injury tax on employers, rather than set standards (see Chapter 8). This is a method favoured by some economists, but not usually by governments (Reppy, 1979). It may provide an incentive to large or wealthy firms to pollute, and the level of fines payable for non-compliance with safety standards might similarly be regarded. It has been argued, for example, that asbestos processors pay much less if fined for breaking the asbestos industry regulations than the cost of installing safer equipment, proper ventilation and adequate protective clothing (Dalton, 1979).

Regulation of the kind discussed above is based on the implicit assumption that a technological solution is possible: for example, that textiles may be produced in such a way that no worker is affected by lung disease. But it may well be the case that some products cannot be made without hazard to the users, the workers producing them or the environment. For example, the Swedish Government (Sweden, 1981) has decided that any use of asbestos is hazardous, and all forms of the material have been banned. An alternative approach is to accept the existence of risk and to measure it explicitly.

9.9   ASSESSMENT OF RISK IN THE CONTROL OF TECHNOLOGY: THE PHILOSOPHY BEHIND REGULATIONS

Risk is usually expressed in terms of the damage that would be done if a particular event occurred multiplied by the probability of that event occurring. (CSS, 1977). However, risk assessment is also complicated by the fact that certain technological changes have benefits to some people, but costs or potential costs to others. In these cases, or even where both risks and benefits are experienced by the same person, some trade-off has to be established: acceptance of a certain level of risk in exchange for a benefit. Where the risks and benefits are experienced by different people, vested interest may be at stake, the status of scientific knowledge may be challenged and in some cases, public costs may be set against private benefit. These considerations are discussed in this section.

First of all, the degree of benefit and risk varies with the individual, apart from any conflict of interests. The case of the oral contraceptive illustrates this. A woman must judge whether the benefits of her not conceiving justify the risk of side effects. But her susceptibility to side effects will vary with her circumstances – for example, smokers and over-35s are more likely to get thrombo-embolic diseases – and risk of pregnancy will also be more or less serious depending on her medical, social and economic circumstances. What is acceptable as a risk varies with the benefit – but also with other circumstances determining the nature of the risk (Walsh, 1980).

Public tolerance of acceptable risks varies a great deal and must be taken into account in drawing up regulations. Thus, for example, public acceptance of risk, and therefore expectation of safety standards, is very different in the cases of nuclear power generation, air travel, and private motor travel. The effect of the media, or public relations campaigns by interested parties, may influence public opinion in relation to risk, but the different standards of acceptability also have quite a logical basis. For example, there may be many more car than plane accidents, but each car accident is likely to kill fewer people. Nuclear power generation has so far produced fewer accidents than other forms of energy, than manufacturing industry, transport and so on, but the damage likely to be done if an accident did occur might be greater. The negative effects of nuclear accidents or unsafe practices such as cancer, are likely to be long term, and are not entirely understood. Thus, individual examples illustrate the wide variation in public acceptability of risk from different products or industrial processes.

In recent years an additional dimension to the acceptability of risk has been the benefit of employment. In a period of high unemployment, workers may well find themselves expected to assess the potential risk to their health of working in a certain environment, relative to the more immediate risk of being unemployed, if they either leave work voluntarily or if their demand for improved safety standards threatens the firm's competitive position and puts all their jobs at risk. For example, a majority of workers at Turner Brothers Asbestos (Turner and Newall) in Rochdale have so far decided the immediate certainty of unemployment is worse than the possibility of disease in the future. TBA's 1982–3 losses made it clear that campaigns for increased protection would threaten the existence of the company, although asbestos workers in other firms have been pressing for a third alternative: diversification into safe substitutes.

Public access to knowledge or information is an important factor in tolerance of risk. For example, people are less likely to accept risks where they feel they lack knowledge. They may lack knowledge because the knowledge does not exist – and further research may reveal new hazards. Or they may lack knowledge because they do not have access to it, as a result of lack of expertise, or as a result of the unwillingness of experts or vested interests to reveal the knowledge. These considerations have been significant in many of the examples of risks given above.

A major problem of risk assessment occurs when the benefits and risks refer to different people or groups. For example, a person may suffer ill-health or lack of amenity as a result of pollution by a local factory, but may not derive any compensating benefit (for example, in the form of employment) from the factory's existence, and may have no opportunity to exercise any choice about whether or how the factory continues to operate. In such cases, assessment of risk and establishment of regulations becomes a matter of achieving an acceptable balance between public costs and private benefit. The government's role is to establish such a balance. In practice, however, it is not an easy matter. Groups with different interests are likely to conflict, and the outcome may be resolved on the basis of the power of the groups concerned, rather than abstract justice. The development of conflicting interests also influences the status of scientific and technical knowledge. Different interest groups frequently attempt to use experts and scientific information to back up their case, thus undermining the status of such knowledge claims generally.

The idea of a trade-off between costs and benefits, and especially between costs experienced by the public and benefits by private individuals, is reflected particularly clearly in British regulations in the key concept of 'best practicable means' (b.p.m.) originally developed by the Alkali Inspectorate. The b.p.m. is what can be agreed between, say, a polluter and the factory inspector as the best level of effluent output, taking into account the economic circumstances of the polluter as well as all the technical means available. For example, a factory inspector may be persuaded that a particular emission level is satisfactory even though it could, technically, be reduced, because further reduction would cost such a lot of money that the firm's profitability or even survival might be threatened. The contrast with the much more openly adversarial process typical of the US is discussed below (Section 9.14), together with the advantages and disadvantages of each.

Irwin's (1985) study of risk assessment and road safety also suggests that approaches to risk mirror national cultural differences. In litigious America, where consumer groups have demanded corporate responsibility for product safety, the major advances in motor vehicle safety have been in occupant protection. In Britain the emphasis has been more on traffic engineering, that is maximising safety through improved lighting, signals and road surfaces.

In summary, it is clear that risk assessment is not a scientific precursor to regulatory policy, but is itself part of the process which shapes policy.

9.10   INSTITUTIONS FOR CONTROL OF TECHNOLOGY IN THE UK

This section describes how the UK Government is alerted to potential hazards that may need to be regulated, how it brings about regulation, and how the regulations are implemented, with particular reference to the plethora of expert committees, licensing bodies and inspectorates that have developed to perform these functions or advise the government in doing so.

It has often been noted that committees are to be found at every level and every turn in British government. *Government by Committee* is the title of a book on the British Constitution (Wheare, 1955). 'There are committees to advise, inquire, negotiate, legislate, administer, scrutinise and control' (Williams, Roy and Walsh, 1982). Science and technology regulation is no exception. Committees are either set up to examine a particular area, for example occupational health, with a view to identifying problems and recommending action by government, or they are established on an on-going basis in order to forecast potential problem areas and act as an 'early-warning' system. Most technologies with potential dangers are under scrutiny by a committee of some kind. For example, the Genetic Manipulation Advisory Group (GMAG) was set up in 1976 to monitor research proposals and identify possible hazards in the field of genetic engineering (GMAG, 1978). The risks and benefits of commercial nuclear power were assessed in the sixth report of the Royal Commission on Environmental Pollution (1976). Regulations to control *in-vitro* fertilisation and embryo experiments were proposed by the Warnock Committee (Committee of Inquiry into Human Fertilization and Embryology 1984). There are many other examples too numerous to mention here, some of which combine regulatory and promotion activities in their briefs (for example ACARD – the Advisory Committee on Applied Research and Development).

The proliferation of advisory committees is evidence of the essentially *ad hoc* and piecemeal approach to the control of technology to be found in Britain. The three main areas in which regulatory actions have been carried out – product safety, health and safety at work and the environment – have all first been subjected to the deliberations of committees. The Government may then do one of four things:

1. introduce legislation based directly on the report;
2. publish a white paper – announcing its decisions and giving reasons;
3. publish a green paper, as a basis for discussion or public debate before committing itself to a specific course;
4. do nothing.

If legislation is introduced in which certain standards are set or regulations established, it is usual at the same time to provide the means to ensure compliance with the legislation. For example, an inspectorate may be established whose members visit work places and other premises to establish that regulations controlling environmental pollution or safe working practices are being carried out; they may advise manufacturers or employers, and if necessary they may prosecute for non-compliance. The Factory Inspectors, Alkali Inspectors and Nuclear Installations Inspectors all have this function. To take another example, a body may be established to issue licenses to sell or manufacture certain products once it is satisfied of the products' safety. This is the case, for instance, with pharmaceuticals and food additives. These characteristics of regulation in Britain can be illustrated by the examples of occupational safety, pollution control and product safety.

In the first case, the long history of state involvement in occupational safety had led to a collection of regulations which many considered to be in need of consolidation. Following the 1970 Robens Committee of Enquiry, the Health and Safety at Work Act (1974) and the Health and Safety Commission (and an Executive arm) were set up. The Act is an enabling act which allows codes of practice to be drawn up for individual industries and amended as appropriate. The Commission has employer, trade union, local government and safety organisation representation, and the Executive organises the previously separate Inspectorates for Factories, for Mines, for Nuclear Installations, and so on. Thus the *ad hoc* regulation of the past has to some extent been supplanted by the creation of an institution with some autonomy and influence.

It is important to note, however, that the original philosophy of the Robens report and the 1974 Act was one which saw safety as an issue for discussion between interested parties rather than collective bargaining between workers and employers. But this philosophy frequently breaks down if one side or the other dispute the adequacy of the regulations or who has been at fault when they are breached in particular cases. Furthermore, as was noted in the previous section, disputes can arise over scientific and technical evidence used to support standards. A clear example is the case of the setting of the 'threshold limit value' (concentration of a substance to which workers may be repeatedly exposed safely) for vinyl chloride monomer which is used in the manufacture of PVC. A tripartite committee recommended a value for the TLV for this chemical which workers in the industry were originally prepared to accept, but when they obtained evidence from the US that this level may not be safe after all, they conducted a prolonged campaign for a lower TLV, using the same approach which might be used for bargaining over pay or hours of work. (Green, 1982).

In the case of environmental pollution, the key piece of legislation, the 1974 Control of Pollution Act, followed a 1970 white paper from the Department of Environment. Like the Health and Safety Act, this is an enabling act which gives local authorities and other public bodies responsibility to secure compliance with standards which they accept as 'reasonably practicable'. With the exception of some sixty 'scheduled processes' which are the subject of more stringent national standards policed by the HSE, pollution is therefore governed through the mechanism of agreement between inspectors and employers over 'best practicable means'. This mechanism has not always been seen as capable of achieving satisfactory results. In 1975 the Royal Commission on Environmental Pollution roundly criticised the inspectorates for being 'remote and autocratic', and for not responding to changing public attitudes to pollution, requiring them to apply more stringent standards (Williams *et al.*, 1982). In the case of water pollution, sewage disposal near public beaches led to threats of the UK government being taken to the European Court of Justice for failing to reach EEC standards, but only as a result of the attention of the media and public pressure, (Pearce, 1984; *New Scientist*, 1985c). As in the case of risk assessment therefore, the roles of information and public attitudes can be seen to be direct influences on the actual effects of the bare instruments of legislation and surveillance.

The case of product safety is complicated by the existence of widely differing product types, ranging from motor vehicles to children's toys. The case of drugs is, however, an interesting example. The key piece of legislation is again an enabling Act, the Medicines Act of 1968. This followed the establishment of the Dunlop Committee on Drug Safety, which was set up following the thalidomide incident in 1964 and the Sainsbury enquiry into the relationship between the drug industry and the National Health Service. The Act gives powers to the Medicines Commission (within the Department of Health and Social Security) to license drugs, to control standards of advertising and marketing, and to oversee voluntary price regulation schemes. The Commission is advised by the Committee on Safety of Medicines which is made up of experts such as professors of pharmacy and senior doctors who act in a part-time advisory capacity. This committee has to rely on published academic data concerning the effects of drugs and (more often) on the submissions of the manufacturers themselves. The main difficulties and controversies which arise are therefore often focused on the quality and origins of this information, illustrating again the possibilities for the regulation mechanism to be sometimes susceptible to conflictual rather than consensual influences.

It is interesting to note that in all three of these fields of regulation the setting of regulatory standards sometimes also has a promotion effect. Compliance with standards often requires innovation which may have favourable results for companies and for exports and growth. For example, a UK flooring company won a Queen's innovation award for a special anti-static floor for hospital operating theatres (Seal, 1968). More recently, in order to comply with regulations limiting the use of asbestos in domestic flooring products, the same company developed a vinyl floor covering that had superior 'lay flat – stay flat' properties as well as no asbestos content (Walsh and Roy, 1983). This example further illustrates the delicate interconnections which often exist in technology policy.

In the case of the UK the overall tendency of the past two decades can be summarised as involving some considerable consolidation of regulatory activity in institutions which have some increased powers, and which are characterised by a consensus-based approach. The flexible character of the enabling acts and the close relations between inspectorates and industries do allow significant power and reach to these organisations. However, there are frequently criticisms of the approach on the grounds of lack of direct public accountability. According to this view, where there is a conflict of interest between industry and another section of the public (for example, the workforce, the consumers) inspectors and licensing authorities may be more strongly influenced by those with whom they work most closely. Thus, for example, the Ombudsman was extremely critical of the Factory Inspectorate in relation to the Acre Mill asbestos plant at Hebden Bridge in Yorkshire (Parliamentary Commission for Administration, 1976), while Social Audit (1974) ('a public interest research group') suggested in relation to the Alkali Inspectorate that 'whatever safeguards do exist for the public in the "Best Practicable Means" they are always liable to be lost in the Inspectorate's overriding concern for the economic consequences of its requirements'. The question is therefore one of the relative power of firms and regulatory bodies.

The contrast between the UK and the US is interesting here. In world-wide terms US firms are perhaps more powerful and influential than any other, and of course have access to a wide range of resources including research, development, testing and the technical and legal advice of experts. But it is also true to say that decision-making in the USA on regulation of technology is also more open and political than it is in Britain. It is taken far more for granted that different interest groups present alternative cases, backed up by expert opinion and often by research. 'In Britain, policy makers adopt the closed consensual and paternalistic approach that proceeds behind tightly closed doors where even pressure groups have to be "establishment" before they can take part.' (Ford, 1985). In the US, it is more usual for the adversarial process to result in the establishment of national or at least state-wide standards, for example, for concentrations of toxic substances, rather than that different standards be negotiated for each geographical area or even each work-place.

Some US regulatory authorities have more facilities than their UK counterparts to conduct independent research in their particular area of concern. For example, while the UK licensing of drugs depends on advice from a part-time committee of lay experts, in the US the Food and Drug Administration employs its own full-time professional staff who may do their own tests to supplement data supplied by firms if they think it necessary, and who may also inspect company laboratories and factories.

9.11   THE EFFECT OF REGULATION ON INNOVATION

Beginning in the US in the 1970s, and increasingly in other advanced countries concern has been expressed about the possible adverse effects of regulation on innovation. A number of academic studies have been carried out, particularly focusing on three of the most 'regulated' industries: pharmaceuticals, chemicals and motor vehicles. There is no doubt, however, that the debate is very much a reflection of political ideology in a period of economic recession, with both 'sides' claiming support of academic studies. Ford (1985), for example, who is himself a politician and former technology policy researcher, has reviewed recent publications in terms of 'the risk analysis of the New Right' versus 'a perspective influenced by the marxoid "radical science" movement'.

The arguments against the regulations are, essentially, that they are causing financial problems to industry and that innovation is being inhibited either by the regulations themselves or as a result of the financial constraints imposed. Clearly, where a limited budget for R & D or for investment exists, regulations may well cause finance to be diverted, for example, from product R & D to process R & D, to secure compliance. Regulations, or the cost of complying with them are likely also to control entry to the market.

However, Ashford *et al.* (1979) argue that the positive effects of regulation are likely to be as great as the negative ones. Possible benefits include new uses for existing products, energy-saving or materials-saving process innovations, and sometimes efficient reorganisation of the firm. We have also seen in the previous section that firms may often capitalise on new market needs and demands created by regulation.

In the case of drug regulations, some authors have described not only an economic cost, but also a social cost – in terms of illness and death that might have been prevented were it not for the delay in the introducing new drugs. Peltzman (1974), for example, did a cost–benefit analysis of the FDA regulations which suggested that it was economically more desirable to have a thalidomide disaster every so often than to have the Kefauver–Harris Amendment to the Food, Drug and Cosmetics Act. Urquhart and Heilmann (1985) were even more forthright in criticising the clinical testing demanded by the FDA. They argue that 100 000 people died who might have been saved by beta-blocker heart drugs during the seven years it took to approve them.

There are in fact several arguments involved in assessing the effect of regulations on innovation. First, were the regulations responsible for the observed decline in rate of introduction of new drugs? Second, if so, was the decline in drugs of therapeutic value, or did the regulations 'weed out' the drugs of little value? Third, has the delay between discovery and launch resulted in more deaths and suffering than would have existed if the tests were not there to filter out drugs with harmful side effects? And fourth, if there is any unduly long delay between discovery and launch, is it due to the regulations or some other reason?

At the Kennedy–Nelson hearings on the drug industry in the US Senate, Schmidt (1974) defended the regulatory authority by arguing that the decline was world-wide rather than confined to the USA; that the decline in the USA was concentrated in drugs of little or no therapeutic value (one of the intentions of the legislation); and that innovation in pharmaceuticals in the 1980s would be limited by the limited range of technical possibilities by then available. In the view of Steward (1978), even though 'in commercial terms, a high rate of innovation may be desirable and a decline is to be deplored, in terms of broader health considerations, more important is the therapeutic significance and character of innovation rather than simply the rate'. A similar argument might apply to innovation in any field.

Grabowski (1976) suggests that it is not so much the regulations, but the bureaucratic way in which they are implemented, which has caused much of the delay. On the other hand, the 'super-cautious FDA' may have delayed heart patients' access to beta-blockers, but they also hesitated to give a licence to thalidomide or Practolol for what turned out to be very sound reasons. They could thus be said to have prevented many cases of suffering and death as a result of the side effects of these drugs, which the licensing authorities in Britain and other countries failed to prevent. Grasham (1985) says of Opren in the UK: 'there was a comprehensive failure by the Department of Health and Committee on Safety of Medicines to give the public the protection it needed and deserved'. Lesser (1983) blamed the magnitude of the social costs on the 'cosy partnership' between the industry and the Committee on Safety of Medicines which 'leads to complete paralysis of action' in a crisis. Urquhart and Heilmann, however, argue (1985) that the deaths from the delay of beta-blockers were greater than for all known drug-induced disasters. No doubt the argument over costs and benefits of drug regulations will continue for some time.

Similar debates have been taking place in relation to regulations and the chemical industry (for example Rothwell and Walsh, 1979) and the motor vehicle industry (Stubbs, 1979; Irwin, 1985). In the 1980s the US and UK governments have both amended some of the regulations governing the effects of technology, employing the argument that the regulations weakened the competitive or innovative position of the industry in question. Nevertheless, some of the underlying pressures toward regulation remain strong, and harmonisation of EEC regulations may bring about a 'levelling-up' effect in Europe.

9.12 CONCLUSIONS

In this chapter we have attempted to give an overview of government policies to promote and regulate technology in the post-Second World War period. Science and technology policy for stimulation of technical change has evolved under the impetus of two factors: the different views of the parties in government, and the increased understanding as a result of research on the innovation process and the relationship between science, technology, innovation and growth. The views of the parties and the results of the research have also been influenced by pragmatism in the face of industry's reluctance to comply with government's wishes (and the pressure of other policy goals such as the level of employment). Regulation of technology has evolved in response to growing public awareness of the consequences of technical change and a desire for greater public accountability by government and industry. It has also been affected by the different approaches of the parties, changing economic circumstances, and increased understanding of the innovation process and of the relevant technologies involved.

10 Non-Governmental Influences on Technical Change

10.1 INTRODUCTION

So far in the book we have discussed technical change carried out by industrial firms under the influence of the market, government policies and the firms' own goals, strategies and general 'culture'. Since approximately 85 per cent of resources devoted to R & D in Britain and America is spent on work done in industrial or government (chiefly industrial) establishments, while over 90 per cent of these funds come from industry or government (Tables 8.1 and 8.2), clearly these institutions play the dominant role in determining the rate and direction of technical change. Individuals who are neither senior managers nor government ministers generally exert an influence over the rate, direction, scale and consequences of innovation only in their role as consumers making choices in the market place or, indirectly, as citizens electing representatives who in turn may determine the policy of government.

The consensus of opinion, at least in countries with market economies and democratically elected governments, is that this is an acceptable, rational and efficient arrangement. Occasionally, however, substantial sectors of public opinion become convinced that the combination of market forces and government policy has been inadequate. This was the case in the 1970s, particularly in the United States and Europe, when widespread public concern about such unwanted effects of technical change as environmental pollution, unsafe products or work hazards obliged governments to tighten up their legislation on these issues (Chapters 8 and 9). Neither market forces nor existing laws were regarded as adequate to protect the public from the unwanted consequences of technical change. The agent of change was that rather hard-to-define entity 'public opinion' reflected in, and reinforced by, the mass media and the campaigns of pressure groups and political parties.

Clearly, therefore, there are institutions partially or wholly independent of firms, markets and governments, which under certain conditions may exert an influence on the rate and direction of technical change or its social and economic consequences. The recognition of these influences on technical change raises further questions, such as: When and how does dissent from the views of firms and governments become 'a shift in public opinion' to be taken into account by firms and governments, rather than the opinions of minorities, cranks or even 'subversives'? Can individuals influence decisions about technology? And should there be more public involvement in controlling or influencing technology, and if so how can it be achieved? In this chapter we shall examine the issue of participation in technological decision-making and consider the role of institutions such as trade unions, and special interest or 'pressure' groups in establishing the rate, direction and consequences of technical change.

10.2 PUBLIC PARTICIPATION IN TECHNOLOGY POLICY

In Chapter 9 we described the changing public view of technology between the 1950s and the 1970s. A generally unquestioning acceptance of technological advance as beneficial and progressive was gradually tempered by concern. Technology was not only seen as responsible for the many ways in which people could exert *more* control over their environment and their lives, but also responsible for many problems, unwanted consequences and an increased *lack* of control. Critics pointed to pollution, loss of privacy as a result of data banks, unemployment resulting from automation, the depletion and possible exhaustion of natural resources and the threat of nuclear war as examples. A sense of alienation as a consequence of technology was commented on by several authors. Thus Freeman (1982) reports a feeling:

> that the whole system is like an uncontrollable and unpredictable juggernaut which is sweeping human society along in its wake. Instead of technology serving human needs it sometimes

seems to be the other way about . . . As a result, the social mechanisms by which we monitor and control the direction and pace of technical change are one of the most critical problems of contemporary politics.

Cooley (1972) related alienation particularly to an increased automation of manufacturing and design. He further argued that an increased work tempo and shift working for some workers, with greater unemployment among others, rather than increased leisure for all, had led to demands by the workforce for more say in decisions made about the adoption of new technologies.

This concern about alienation was reflected in the debate about increased participation in decision-making on technological issues that took place during this period. Benn, for example, after a number of years as Minister of Technology, argued in favour of increasing the role of the public in establishing priorities in various areas of government expenditure. It was, he said, only the detailed process of implementation which was technically complex and involved experts in the decision-making: in establishing broad priorities, the decision was similar to other choices put before the electorate, even though the subject involved technology (Benn, 1971). When he was back in office, Benn tried to put some of his views on more 'open government' into practice in his capacity as Energy Secretary and, for example, organised a one-day public debate on nuclear fuel reprocessing at Windscale in January 1976.

An alternative position was put forward by Williams (1973), who argued against public *participation* on the grounds that the need for expertise, the constraints of time and the need to protect the interests of people who do not take part in decision-making, made it impractical. However, he did favour the increased *accountability* of public bodies; that is, that public officials and institutions should be more open to investigation by members of the public, and obliged to account for their actions, to ensure they acted in a responsible way. However, successive UK governments (of either party) have been reluctant to concede that even the degree of accountability required of US public institutions is desirable in the wider public interest, as is shown by the regular recurrence of debate about the relative merits of the US Freedom of Information Act compared to the UK Official Secrets Act, both of which cover technological as well as other kinds of decisions.

At an official level, public involvement in decision-making has been very limited. In the early 1970s some experiments were conducted in various countries including F. R. Germany and Sweden, whereby large numbers of the public were invited to comment on technological issues (such as nuclear power) via the media, radio phone-in, and public meetings organised in conjunction with community and religious groups. In Britain Benn's support for public involvement in decision-making led eventually to the 'public enquiry into the planning application by British Nuclear Fuels Limited (BNFL) to construct an oxide fuel reprocessing plant at Windscale' in the words of Environment Secretary Peter Shore (1977): The Windscale Enquiry. The UK Town and Country Planning Act (1971) requires local planning authorities to make their proposals public, and if necessary hold a public inquiry if serious objections emerge. The Windscale Enquiry was held under the provision of this Act.

Giving evidence to a public enquiry is one way in which members of the public may alter decisions with which they do not agree. It is, however, fairly limited. Public enquiries do not happen very often, and most members of the public, even if they feel very strongly about an issue, do not feel they have the time or the expertise to present evidence at a public enquiry, or the resources to commission an expert to do it for them.

In the US, the National Environmental Protection Act (1970) requires federal government agencies to file environmental impact statements, which analyse the possible impacts of proposed projects. These statements (described in Section 9.8) are made publicly available and have led to many court cases (five hundred in the first five years) instituted by citizens who objected to the proposed development. The Office of Technology Assessment (also described in Section 9.8) was established to take a longer term view, and assess future direct and indirect effects of technological change in both public and private sectors. The OTA initiated a number of experiments in public participation in their assessments, including questionnaires, surveys, and the establishment of panels of 'citizen representatives'.

According to Nelkin (1977), there are several areas in which problems must be resolved before public participation in decision-making can be at all effective. First of all, the right balance needs to be established between the fullest possible participation and the most efficient means of decision-making. Then, it needs to be decided who should be involved. If all interests are to be presented when decisions are made, some acceptable definition of what is a legitimate interest, and who is representative of the interest, needs to be established. If participation is not to be just a rubber-stamping procedure or a reaction to formulated policy, then participation needs to begin at an early stage in the decision-making process. If members of the public are to be fully involved in decisions affecting technology, then there must be some means of improving the public's understanding of science and technology. Lastly, it must be recognised that political conflicts cannot be ignored. They really exist, as reflected in a number of controversies over technology which have involved political values and priorities as well as technical issues. We shall return to the question of conflict and consensus in Section 10.5.

10.3   THE TRADE UNION MOVEMENT

Usually, the first action taken by ordinary citizens who wish to influence a decision on a technical (or non-technical) matter is to try to get someone with recognised authority to act on their behalf. This, after all, is the basis of representative systems of government and is often an effective course of action. The representative in question might be an MP, Councillor, or trade union official, for example. Another way in which individuals may have some influence over changes in technology (or other kinds of changes or decisions) is to join together with other people who have similar interests or views. A group of people is likely to be more effective than an individual in lobbying elected representatives, and in generating the resources necessary to publicise an issue and research practical alternatives, and is likely to have more bargaining power. This is the basis of trade unionism, political parties and pressure groups. Employees in a particular industry, with a particular skill or a particular employer, for example by acting collectively in a trade union rather than as individuals, have more power to improve wages, hours worked, safety at work, conditions of work and so on. Although influencing the rate, direction and consequences of technical change is not a primary aim of trade unions, they may certainly try to have some effect in this area, particularly where the technology in question directly affects the jobs of their members.

Historically trade unions have not been directly concerned with technological issues. Indeed, many trade unionists have taken the view that technological issues are outside the scope of, or even in conflict with, their primary aim of securing their members' jobs and improving their wages.

Cases where the rate and direction of technical change has had a direct effect on trade union members' jobs, skills, wages, health and safety or the nature of their work, have clearly been the most likely to attract attention from unions, and unions have developed strategies aimed at influencing such changes. However, a number of unions have begun to adopt policies towards technological change or particular technologies, less in an attempt to use their bargaining power as a negative and defensive reaction to a perceived and fairly immediate threat, and more in an attempt to develop a positive, longer term strategy aimed at influencing future trends and the general environment.

Several trade unions have published documents dealing with the likely impact of new technology based on microelectronics innovations for example UCW (Walsh *et al.*, 1980); BIFU (1980), POEU (Darlington, 1979); and GMBATU (1980), and many unions have adopted policies towards the introduction of such technology. Among the possible costs to their members of microelectronics technology, seen by the unions are:

1. A reduction in the number of jobs available, including the possibility of redundancy. For example, automation of 'Mr Kipling' pie manufacture now means that one production line can produce 40 000 pies in an hour, requiring just one worker to operate it – the woman who feeds the foil dispenser (Huws, 1982).
2. A reduction in the skill required by those whose jobs remain. Examples such as computer aided design and diagnostic computers in medicine can make major contributions to efficiency, but at the same time they take away much of the skill involved in the jobs of design and diagnosis respectively, reducing the job satisfaction of the staff concerned.
3. More control by management of the day-to-day organisation of the job, the pace, order and style of work. For instance, a

secretary using a word processor has her work organised to a far greater extent by the equipment. Before automation, she may well have organised her job to maximise the variety of typing, filing, telephoning, taking messages, going to take dictation, sending out mail, and so on.
4. Increased risks to health, such as vision problems allegedly associated with the use of Visual Display Units, or increased stress as a result of the reduced variety or skill involved in certain tasks.

On the benefit side, unions have taken the view that difficult, dangerous, dirty or otherwise unpleasant jobs could be automated to the workers' advantage. Some unions have suggested that improved efficiency, as a result of introduction of new technology, will increase wages and security to their members, while firms which do not modernise may well be forced out of business altogether. In some cases the opposition and resistance to new technology is not so much an opposition to the equipment itself, but workers' frustration at the lack of control or influence over decisions relating to their jobs.

Before the widespread introduction of microprocessor-based technology, other forms of automation had been introduced. In many cases these innovations have also led to resistance by the workforce or to attempts to influence in some way the extent or timing of the change or the conditions under which it was introduced. Similarly, technical changes potentially affecting the health and safety of employees such as the introduction of a new chemical process using or producing chemicals of unknown toxicity, have also led to attempts by unions to influence decisions or to impose conditions about their introduction.

The present climate of economic recession has made trade unionists particularly aware of the vulnerability of firms, and in particular the competitive weakness of many British firms in the international market, and consequently the need for companies to introduce new technology to improve efficiency and to make better products. On the other hand, the very same economic climate has also made trade unionists very conscious of the scale of unemployment and the difficulty of finding new jobs in the event of redundancy.

The result has been a qualified support for the introduction of new technology, the qualification being the use of collective bargaining and other means to influence the distribution of the benefits of technical change. Several trade unions, for example, have campaigned either nationally or in particular work places for a shorter working week in the event of new technology being introduced. In some cases ambitious proposals have been put forward by the workforce of a particular company or group.

One of the most well-publicised of these was the Lucas Aerospace Joint Shop Stewards' Alternative Corporate Plan (Wainwright and Elliott, 1982). These workers, who were members of several trade unions, took the decision to try and take a positive stand on an issue related to technology and, rather than just resist change, to make concrete proposals for diversification of production. On several occasions in the period 1974–85 Lucas Aerospace was faced with the necessity of rationalisation and cut-backs, first as a result of the Labour Government's defence cuts, and more recently as changing technology led to changes in demand. For example, in 1982 aeroplane design changes by Lockheed led to aero engine design changes by Rolls Royce which, in turn, meant the cancellation of orders for some components from Lucas (Hildrew, 1982). On each of these occasions, the workforce opposed redundancy and proposed diversification. They were quite specific in their proposals and the Alternative Corporate Plan consisted of detailed designs and specifications for products, suited to the company's production plant and equipment and the skills of the Lucas workforce, and for which they had established there was definite demand.

However, the approach by the Lucas trade unionists met considerable resistance from management, not over the technical or marketing details of the proposals, but over the existence of such proposals at all. Essentially, what was at stake was management's prerogative to manage and make decisions. On management's side were arguments about the inefficiency of involving large and variable numbers of people in decision-making, the possible questioning of their own authority and status, and the feeling that workers who were encroaching on management's job were not doing their own job adequately. The workforce, for their part, took the view that corporate strategy affected them directly; since they worked for Lucas their interests were bound up with the future of the company and so they should have some right to a say in the decisions that were made, especially if they were making positive proposals and not just reacting defensively and negatively to change.

Similar alternative plans have been discussed by workers faced with redundancy at Vickers, British Leyland, Chrysler, Clarke Chapman, C. A. Parsons, GEC, Rolls Royce, Hawker Siddeley and British Aerospace (Wainwright and Elliott, 1982). The Lucas plan however, remains the most detailed example of a 'workers' alternative corporate plan' as well as the most well known. Much of the emphasis of these plans was on the diversification into 'socially useful' or 'environmentally appropriate' products and production methods. However, there are many contrasting views within the trade union movement on these issues, and strong pressures still predominate towards seeking short term advantages rather than longer term social and environmental benefits. Nevertheless, many trade unionists have begun to question the nature and purpose of their work, as well as their wages and security, especially in the context of advanced technology.

The normal mechanism through which trade unionists may attempt to raise such issues is collective bargaining. Throughout the 1970s collective bargaining gradually extended beyond the issue of wages to issues concerning the organisation of work and the allocation of resources. During the same period, 'industrial democracy' was very much in vogue, though meaning widely different things to different people. Several firms initiated worker participation schemes, or joint consultation, where details of work organisation, including the manner of introducing new technology, could be discussed. Concern about alienation of the workforce as a result of repetitive or boring tasks or loss of skill due to automation led to various 'job enrichment' schemes. For example, a particular job might get added responsibilities such as quality control or coordinating material flow, to increase the workers' interest in their jobs.

Such initiatives were rarely welcomed by the unions. Job enrichment was seen as very limited, constrained by production targets still set by management and in some cases a form of 'back door' productivity deal. Consultation was seen as an erosion of trade union power; some of the issues had previously been the subject of collective bargaining, in which the trade unions felt they had some power, but in consultation there was no obligation on management to accept suggestions from the workforce.

The 1974–9 Labour Government attempted to institutionalise 'industrial democracy' through the provisions of the Bullock Report (Committee of Inquiry on Industrial Democracy 1977) and the planning agreements of the 1975 Industry Bill. These attempts were not overwhelmingly successful in increasing the power of trade unions in decision-making, over matters concerning new technology or anything else (see Chapter 9); and the 1979 Conservative Government has since ended such rights as were established by trade unions to have an official voice in decision-making, by private or publicly owned companies. Trade unions still do, of course, attempt to exert some influence over industrial decisions affecting technical change and other issues, but this is now largely limited to traditional collective bargaining, or to more overt political campaigning.

An example of the latter is the National Union of Agricultural Workers' campaign, begun 1980, to have the weedkiller 2, 4, 5-T banned (Cook and Kaufman, 1982). This began as a health and safety at work issue for the union's members, but rapidly expanded to a campaign for protection of the environment and the general public. The weedkiller, available over a chemist's counter to gardeners or used in bulk by city parks departments and farm workers, was used as Agent Orange, a defoliant, in Vietnam where its toxic, carcinogenic and teratogenic effects were first observed. The campaign rapidly became a political issue taken up by other trade unions and political parties, leading to attempts to get the Government to ban the chemical. So far, however, 2.4.5T has not been banned, and many criticisms have been made by the campaigners of the Government and its Pesticides Advisory Committee. This type of campaigning is essentially an attempt by trade unions to shift the boundary between the government and the market in their relative control over technological matters.

10.4 PRESSURE GROUPS

The example of the trade unions has shown that when a group of citizens, in this case industrial workers, have interests or opinions about a new technology which do not coincide with those of either the government or firms' management they can find another institution to represent their different interests. Neither technology nor social systems are static and changes in both of them can create conflicts of interest that no present institution is capable of handling. Thus many new pressure groups have become increasingly prominent in the political systems of western countries, and a number of them have been concerned with issues with a strong technological dimension.

The Birth Control Movement developed in the 1920s and 1930s, especially in Britain and the US to campaign for legalisation of birth control (in the US) or to make it more widely available. They pressured the government and the medical profession to accept it, they opened clinics, they did quality control tests on existing products, they carried out research on new or improved methods, they provided advice and they publicised birth control. Although the fact that birth control is widely accepted, as well as practised today is not solely the result of campaigns by the birth control movement, but of social, economic and political changes (Walsh, 1980), nevertheless this pressure group achieved a considerable degree of success. Regarded as a radical and even downright disreputable fringe group in the 1930s ('fanatical propagandists and hysterical ladies' according to the medical press, Kennedy, 1970) organisations like the Family Planning Association (UK) or Planned Parenthood (US) are today pillars of respectability, chaired by eminent citizens including a US President or two.

A more recent example of a pressure group which has achieved a certain degree of success is the Consumer Movement. In the UK its main influence has been the provision of information to consumers to enable them to make more informed choices of products within the limits of existing technology and the existing market. Campaigns for Real Ale and Real Bread have had some success in persuading the brewing and baking industry that there was a demand for traditionally made beer and bread. However, the consumer movement in the US has been much more political. Ralph Nader's Public Interest Research Group successfully challenged the US Government and some large corporations over motor car safety, food additives and the withholding of information, inducing changes in the law to improve product safety (see Chapter 9).

The environmental movement grew into a fairly significant lobby in the west in the 1970s, consisting initially of pressure groups opposing various planning proposals (such as those for airports or motorways) and developing to cover pollution by pesticides and other chemicals and the waste from nuclear power generation. Organisations such as Friends of the Earth, the Conservation Society and the Town and Country Planning Association were among those giving evidence against planning permission at the Windscale Inquiry (see Section 10.2). The Inquiry found in favour of the proposals, but it was a measure of the environmentalists' influence by 1978 that the outcome was ever in doubt.

More recently, the campaign to remove lead from petrol has been successful in the US where new cars are now required to run on lead-free petrol. In 1981 the UK authorities decided to keep lead in petrol but organisations such as Campaign Against Lead in Petrol and Campaign for Lead Free Air (Wilson, 1983) have continued to press the government to change its mind. The campaign has produced a large amount of epidemiological and other evidence to support its case and has had a considerable influence on public opinion.

The British Society for Social Responsibility in Science (BSSRS) has acted as a pressure group in a number of cases related to technology. For example, it has carried out research on the occupational health hazards of working with certain materials (for example asbestos, PVC), supplying information to trade unions and publicising the dangers (for example Dalton, 1979; Green 1982). To a large extent it has acted together with trade unions rather than alone, and has tended to find out and supply technical information about controversial technological issues, while the campaigning and negotiating has been more the province of the unions.

The birth control movement, consumer movement, environmental movement, campaign against lead in petrol and BSSRS, mentioned above, are all examples of pressure groups of varying sizes, concerned with a single issue or a range of related issues, which have tried to influence firms and/or governments about the rate, direction and consequences of technical change. They have attempted to shift the boundaries between institutions' relative influence over technologically related decision-making for example, so that government assumes more control relative to firms and markets, or even so that some mechanism is established to increase the direct input from members of the public into the decision-making process of firms or governments.

They were all started by a small number of individuals, battling against established public opinion as well as governments or industrial firms, and have subsequently enjoyed varying degrees of support, public attention and success or failure in their aims of affecting technological decisions. Where they have been successful they have publicised their case and won the support of a sufficient number of people to be seen as representing a substantial sector of public opinion. The environmental movement certainly influenced the decision-making process at Windscale, although it did not win its case. Even in 1971 Shirley Williams was able to say:

The phrase 'environmental lobby' has become part of the vocabulary of politicians. Five years ago it hardly existed. Any MP whose constituency contains an airport, an urban motorway or a main highway cutting through a village or market town has had impressed forcibly on him the cost people pay for jet aircraft, heavy lorries and the speedy movement of cars. The great difference from the recent past is that growing numbers of people are now determined to do something about it (Williams, 1971).

The Birth Control Movement was so successful that it can no longer be considered a pressure group at all. Representing a minority view in the 1930s and vilified in the press at the time, organisations like the Family Planning Association in Britain now represent the overwhelming majority of opinion in British society, to such an extent that it is now the opponents of birth control who could better be described as an embattled minority. The recent House of Lords ruling in the Gillick case (The *Guardian*, 1985) upheld a 1980 DHSS circular which allowed doctors to prescribe contraceptives to girls under the age of sixteen without parental consent, demonstrating the extent to which contraceptive technology is accepted in British society.

Institutions such as trade unions and pressure groups may thus influence the rate, direction and consequences of technical change where the market mechanism and government policy fails to satisfy a substantial sector of public opinion in this respect.

10.5   CONFLICT AND CONSENSUS

As members of the public involve themselves in decision-making on technological issues, either through pressure groups or trade unions, or the development of some kind of scheme for public participation, it becomes clear that a consensus view will not necessarily be the indisputable outcome of an assessment of all relevant facts.

To begin with, facts are not always neutral in the face of conflicting interests. Scientific and technical data are collected on the basis of models and assumptions which may reflect different material interests, political values, judgements and priorities. A pressure group or trade union, in opposing a government department or a firm over a technological issue, may well dispute either the facts themselves, or which ones are relevant to the issue in question. In the United States it is taken for granted that this may happen, and the experts may disagree, while it is much more common in Britain to assume that once all the facts pertinent to a particular decision are available, the outcome cannot be in question. Technology assessment in the US for example, is based on an 'adversarial' approach. That is, a recognised technological opposition (for example to the decision-makers in a particular government department) are expected to produce counter assessments and arguments, backed up by technical data (Collingridge, 1980).

In Britain many government departments responsible for technologically-related decision-making in controversial areas have established independent committees of experts to advise them. For example, the DHSS is advised by Committee on Safety of Medicines and the Health and Safety Executive is advised by the Advisory Committee on Toxic Substances. These committees may be permanent, or established to examine a specific problem. They may include, besides expert professionals who are expected to be independent, representatives from trade unions, industry or local government. The experience of the Advisory Committee on Toxic Substances shows that the assembly of all necessary facts does not always lead to a consensus view of the necessary action to take (Connor, 1984). The Committee has agreed on a code of practise for the control of substances hazardous to health and a number of recommendations for legislation in 1986–7, but has failed to agree on carcinogens. In particular, the trade union and industry representatives are unable to agree on what constitutes a carcinogen – something with carcinogenic properties in animals or only those substances known to be carcinogenic to humans. To a limited extent, the adversarial process has been recognised in that a seven-month 'period of consultation' followed publication of the Committee's findings in August 1984.

In many areas of decision-making in Britain, voluntary organisations, independent researchers and pressure groups have taken on the task of supplying technical and non-technical information to back up an alternative view to the 'consensus' or official one. Thus organisations like Friends of the Earth or the British Society for Social Responsibility in Science have undertaken research on particular issues as well as being involved in pressure group activity like lobbying and organising campaigns. Some community organisations have received funds (for example from the Manpower Services Commission) to carry out research and develop alternative policies on behalf of groups of residents or workers who believe it is necessary to challenge an established view about the safety of a particular building or emissions from a local factory, or the inevitability of the closure of a work place. For example, the Manchester Employment Research Group has prepared an alternative corporate plan in collaboration with the workforce at Leyland Trucks (MERG, 1982), BSSRS supplied a Welsh tenants' association with information about pollution from the local PVC factory, and health and safety information to the trade unions in the PVC factory (Green, 1982). In the 1930s the Birth Control Investigation Committee carried out research on hormones twenty or more years before the pharmaceutical industry began to investigate hormonal contraception (Walsh, 1980).

Many trade unions now have their own research departments whose job is to supply the union's negotiators with technical, economic and other information as an alternative to what is available from the employers. As far as technological issues are concerned, this information is most commonly in the area of health and safety at work, or the effects of new technology on jobs and skills. Several trade union research departments also obtain information from outside experts. For example, the Ruskin College Trade Union Research Unit carries out research on levels of wages and conditions at the request of trade unions. The first occasion when a national trade union officially commissioned research from outsiders on a technological issue was the Union of Post Office Workers' request to the Science Policy Research Unit at Sussex University to provide them with an alternative forecast of the impact of new technology to the one produced by Post Office management (Walsh *et al.*, 1980).

In other countries this sort of adversarial activity is far more established. In the US a range of public interest research groups have been established to provide expertise in the adversarial process of technology assessment. In Denmark the official collaboration between academics and trade unionists, sponsored by public funds, is a regular occurrence. For example Petersen and Venstrup-Nielsen (1984) investigated the hazards to health of mineral wool (an asbestos substitute) used for heat insulation. The research group were based at the Department of Occupational Medicine, Aarhus University and the project was led by a medical doctor and a safety and health officer from the building workers' union. Their findings (including carcinoma in animal studies) were presented to the government department responsible for safety standards in the workplace, to the trade union, and to the manufacturers of the mineral wool. In Holland, 'science shops' have been established at some universities. Technical problems faced by a community group or trade union may be brought to the science shop, which commissions a member or members of the university staff with the necessary expertise to solve the problem. The research is paid for by public funds via the university's overheads.

However, despite these developments, it is still the case that the major problems facing 'public interest' research groups in Britain are lack of funds and access to information. In order to develop a responsible and adequate critique of existing policies or projects, sufficient funds are needed to support research workers. Sometimes this type of work may be supported by charitable foundations or the Manpower Services Commission. In some cases the groups concerned rely entirely on membership subscriptions. The Economic and Social Science Research Council has an 'open door' scheme which may provide funds to do research in, for example, the management and industrial relations area on behalf of a non-university organisation which could be a small firm, trade union or community group.

Information is frequently classified or subject to commercial confidentiality. The Official Secrets Act and Company Law restricts the information available to researchers in this area. This contrasts with the much more open approach of the US Freedom of Information Act which allows access to official information unless it is covered by state security considerations. Access to information is not restricted to US citizens or residents.

10.6 THE ROLE OF EXPERTS

One of the main consequences of non-governmental influences being brought to bear on technological issues is that scientific and technical debates have become much more polarised. Controversies over nuclear power, lead in petrol, safety at work, pollution or the safety of contraceptive and other drugs, have all become essentially political rather than just technical issues. With that development, our perception of science and technology has become radically altered.

All groups of people with a particular interest in a complex conflict associated with technology have come to treat expertise as a resource like other resources available to campaigners (for example, publicity, relations with media, support from well known public figures, legal powers). Whether or not the knowledge or expertise is 'correct' in any absolute sense becomes secondary to the status of knowledge as 'knowledge', which in turn depends on such matters as the status of the expert and the source of the information. Knowledge has become very closely related to power: when all the experts back up a government department, local authority or firm with their scientific information, then those bodies appear to be all-powerful. When some people disagree, and back their argument with expert knowledge that challenges what has previously been accepted as 'the facts', then that power is also challenged. Thus, decision-making in areas which had previously been seen as part of a paradigm of shared rationality and consensus, has begun to take on an adversarial, or negotiated character. In the case of health and safety at work for example, the Robens conception of consensus based on objective scientific facts has given way in the public consciousness to a conflict view, to be settled by negotiation and bargaining like other industrial relations matters, and in which technical information has become part of the adversarial process.

The consequence of such a process has been to weaken the status of scientific knowledge. Once such knowledge is no longer seen as absolutely unquestionable, objective, pure, esoteric and inaccessible to the public, then its status is in some ways devalued together with the political power of the experts themselves. This provides a self-limiting character to the process, and a tendency towards a new equilibrium based on a subtle balance between a strengthened role for 'knowledge as power' on one hand, and a weakened status for particular pieces of knowledge on the other hand. In one sense this increased politicisation of technical knowledge is the most important common residue of the developments within trade unions and pressure groups which have been discussed in this chapter. The contradictory scramble for 'technical evidence' coupled with the scepticism toward the evidence of others constitutes a new dimension to the political processes surrounding technical choices. Thus our understanding of technical change must, where appropriate, take this new source of influence into account. Along with firms, markets and government policies, the consequences of these more or less organised 'public assessments' of technology shape the process of technical change in a complex way, particularly in the case of the more large-scale and visible technical choices. Their outcomes often contribute to changes to the boundaries between firms, markets and state agencies in their relative influence over the technology in question.

11 Summary and Conclusion

Our aim in this book has been to review a substantial part of the literature on technological change and to explore certain ideas which suggest the possibility of convergence and synthesis of previously separate research traditions. It is therefore appropriate for this brief concluding chapter to take the form of a summary of some of the principal themes which have emerged in preceding chapters, and the areas where convergence is most apparent.

Innovation and technical change have long been regarded by all strands of social science as processes which are fundamental to the development and reproduction of industrial societies. Indeed, since Schumpeter, it has been common to argue that development is a necessary feature of reproduction. One of the main obstacles to a more concise description of exactly how technical change performs this central role, has been the uncertainty about how far the process is exogenous or endogenous to the system. Classical economists were intuitively sympathetic to an endogenous mechanism, but did not formalise it. For much of the twentieth century the pendulum swung the other way. Economists regarded technology as exogenous, either because they felt this was accurate, or because this was deemed necessary to simplify the assumptions underlying the other problems with which they were struggling. Eventually however, the institutionalisation of R & D, and the manifest role of science and technology as resource-consumers, together with the increasing role of government, shifted the pendulum back toward a search for endogenous descriptions of technical change.

The search was conducted on a number of fronts. Economists in the neoclassical tradition began to revise their models and to apply formal profit-maximising logic to the behaviour of firms who expended resources on R & D. A growing group of inter-disciplinary researchers examined individual instances of innovation, and began to generate some general statements about the organisational and economic determinants of the innovation

273

process, as well as observing some patterns amongst samples of innovations. Other economists became concerned with the consequences of technical change for growth in output, for employment and for trade patterns. Finally, political scientists and policy researchers became concerned with the formation of sensible guidelines for the corporate and governmental investors and regulators who have become increasingly involved in the complex decisions surrounding technical change.

In order to confer some endogeneity on technical change the first step must be to regard it as a feature of the rational behaviour of business firms rather than as solely determined by the institutions of science, assisted by injections of 'entrepreneurship'. Thus the issue of a theory of firm behaviour is raised immediately. Managerial theories (together with behavioural theories) have proved increasingly popular as test-beds for the most recent generation of technical change specialists. But the analytical approach derived from the production function and the neoclassical theory of the firm have provided important insights in such fields as induced innovation.

The main thrust of those writing within a broadly managerial or behavioural paradigm, has been to present firm behaviour as a function of the growth aspirations of groups of managers, subject to various constraints. Technology enters this picture as a source of both constraints and opportunities. Available technologies render various strategies more or less costly, more or less rewarding, and more or less risky. These considerations are, of course, closely intertwined with the other types of factors which take a part in the formation of business strategies, such as rivalry, the possibilities of new markets, the possibilities for integration or dis-integration and so on.

The technical dimensions of business strategies can be analysed in a variety of ways. In this book we have argued that the R & D activity of firms and of groups of firms can be usefully analysed in terms of the dimensions of size, degree of radicalness, degree of diversity of directions, and finally in terms of the specific performance objectives. These dimensions of technical strategy are to be seen in the light of other dimensions of strategy concerned with marketing or production goals (or other less specific goals). Furthermore the dimensions of technical strategy can then be shown to be related to specific characteristics of the firms themselves which often result from the accumulated experience of the management team. It is against this background of structural analysis of the R & D activities of firms that the variables of technological opportunity, demand, and factor costs can be deployed to explain the inducement mechanisms which favour some patterns of innovation over others.

A major consequence of this analysis is that technological opportunities are to some extent, despite scope for individual firm differences, *common* to some groups of firms. This suggests that there will be some structure to the pattern of innovations generated at the industry level. This issue was examined in Part II of the book.

The patterns to be found among innovations were seen to be an influential factor in a number of partially discrete research areas which have been concerned with technical change over the past two or three decades. The technology-push/demand-pull debate appears to be partly resolvable in several branches of the chemical industry if the patterns of innovation are examined over a long enough period. Induced innovation theories have made progress at the microeconomic level by incorporating the notion of a structured set of technical possibilities. The debate on innovation and market structure has moved from considering how market structure might influence innovation, to considering the possibility that there may be a significant causality in the opposite direction. The notion of natural trajectories of innovation in which progress is channelled in certain directions by a combination of technical and economic forces has proved both appealing and useful, even to the extent of being incorporated in formal growth models. Finally, the new models of diffusion developed in the last ten years have combined several of the above results by incorporating the notion of post-innovation improvements and a changing population of adopters and suppliers responding to changing incentives.

A common thread in this work is its tendency to strengthen the position first associated with Kuznets and Salter which argues that particular industries have specific characteristic rates of technical change which differ from one another, and which evolve over time. Indeed, Nelson and Winter's contribution to the microeconomics of technical change starts to some extent from a concern to identify the roots of the 'productivity puzzle' in differing intrinsic rates of industry technical change. When these ideas are set in the context of a macroeconomic model such as that of Pasinetti, in which sectoral rates of productivity growth and demand growth are independent of each other, the radical result is that equilibrium growth and full employment will be the exception rather than the rule.

Natural trajectories and various kinds of technological paradigms evolve in the course of time. Common patterns of techno-economic behaviour are gradually adopted by different firms even when they start from different initial technologies. However, the process of convergence of different firms' technologies has to be placed in the context of national and international economic development. For example, are the life cycles of different technologies independent of one another, or are they related in such a way that the regularities in the development of individual technologies can produce regular and discrete patterns at higher levels of aggregation in the economy? This type of problem is addressed by the Long Wave literature which suggests that major shifts in technical infrastructure of the economy are not 'digestible' other than as a series of well-spaced structural changes in the mix of industries and patterns of demand. The spacing mechanism is partly economic in the strict sense that the exploitation of one set of technical possibilities affects the incentives to develop and exploit the new, and it is partly institutional in that the major structural changes may have profound social and political implications.

Therefore, once we allow a specific structure to the range of technological opportunities into the picture, rather than a general notion of undifferentiated technical change 'falling from heaven', we find that it has profound consequences for economic analysis at the level of the firm, the industry, and the economy as a whole. The direction of firm growth becomes subject to technical influence, industry structure and productivity are also conditioned by technology, and even the long run patterns of output and employment of national economies are affected. This is not to say that all these matters become 'determined' by technology in a strictly mechanical sense, but simply that some options are ruled out, and of those that remain, some are more profitable, and more likely for some firms than others. What is ruled out or considered largely improbable is the possibility that changes in the environment in which sufficiently mature firms and technologies operate can lead to drastic *qualitative* changes in the firms or technologies themselves. The role of demand, and of second and third order effects operating through price mechanisms is not abolished by these assumptions: rather it is accomodated by *quantitative* changes which do not alter the fundamental nature of the techno-economic system which had previously evolved. Consequently, despite the important role of firms and markets as the central foci of much economic analysis, the arguments of this book suggest that the technological level of aggregation may be an important complement to the industry level of aggregation, despite the difficulties which still exist in applying it.

The above arguments, which derive from the first two 'economic' parts of the book, suggest some conclusions concerning the success of the traditional neoclassical economic paradigm in analysing technical change. There are several instances where an analytical approach which assumes profit-maximising subjectively rational behaviour has given useful and interesting results, particularly at the microeconomic level. Examples are the work of Stoneman, Dasgupta and Stiglitz, Binswanger, and others. Overall however, these contributions do not show the promise of synthesis into a satisfactory theory of the complete process of technical change. So many of the *ceteris paribus* assumptions have to be relaxed in order to introduce the realities of technical change, that the production function approach quickly runs into insuperable difficulties. We have consequently found that various alternative paradigms have been essential sources of the main theoretical advances in analysing technical change. Managerial theories of the firm, Schumpeterian approaches to competition and rivalry, and evolutionary theories of industrial and economic growth have proved to be more fruitful frameworks for the integrated discussion of technical and economic phenomena. This judgement may not find agreement amongst admirers of the neoclassical paradigm, but it does reflect the balance of the interesting and influential literature on technical change in the past fifteen or so years.

But the structure of technical opportunities is not only the result of the R & D process and the operations of the selection environments, rivalry mechanisms, and price mechanisms. The role of government and other sources of social and political influence on resource allocation to R & D, and on utilisation of its outputs, have been shown to be significant and increasing. Governments defend and promote the integrity and prosperity of nation-states through a combination of economic, industrial, environmental and military policies which have an increasingly explicit technological dimension. In the formulation of technology policy, although specific industry and interest groups form two important reference points, it is increasingly the case that specific technologies (information technology, for example) are the objects of policy. This lends further weight to the importance of the technical level of aggregation in social science analysis.

As well as intervening in specific fields of technology, the role of government in trading-off equity and efficiency involves an element of regulating the externalities of technical change, and of dealing with areas where markets and firms are not considered to be performing all the activities that society finds necessary. These activities are closely entangled with the frequent intervention of governments in the process of drawing boundaries between firms and markets, between firms and firms, between firms, markets and public organisations. These decisions are particularly sensitive since they are conducted against the background of the historical argument in western democracies over the relative merits of different levels of 'interference' with the operation of markets, and of the evolution of the industrial corporation as the main organ of coordination of resources. The trans-national character of both markets and firms and their consequent ability to escape more easily the actions of governments adds a new dimension to this argument, both in general, and in the specific context of technical change.

Finally, if the long wave arguments are correct, governments also bear some of the responsibility for initiating some of the institutional innovations which might facilitate the shift from one technical paradigm to another at the lower turning point of the long wave. This may be the most fundamental sense in which development is a precondition for reproduction. However, the economic circumstances of the long wave make the trade-off of equity and efficiency more difficult rather than more easy, thus reducing the scope of action of governments. Perhaps necessity will, after all, be the mother of invention.

Bibliography

Abernathy, W. J. and J. M. Utterback (1975) 'A Dynamic Model of Process and Product Innovation', *Omega*, III, 6.

Allen, G. (1976) *The British Disease: A Short Essay on the Nation's Lagging Wealth* (London: Institute of Economic Affairs).

Ansoff, H. J. and J. M. Stewart (1967) 'Strategies for a Technology-based Business', *Harvard Business Review* (November/December).

Arrow, K. (1962) 'The Economic Implications of Learning by Doing', *Review of Economic Studies*, XXIX, 80, 155–73.

Ashford, N. A., G. R. Heaton Jr. and W. C. Priest (1979) 'Environmental, Health and Safety Regulation and Technological Innovation', in C. T. Hill and J. M. Utterback (eds) *Technological Innovation for a Dynamic Economy* (Oxford: Press) Pergamon.

Baker, N. 'R & D Project Selection Models: An Assessment', *IEEE Transactions on Engineering Management* pp. 165–71.

Balachandra, R. and J. Radin (1980) 'How to Decide When to Abandon a Project', *Research Management* XXIII, p. 4.

Barrett-Brown, M. (1958) 'The Insiders', *Universities and Left Review* (Winter).

Bastos Tigre, P. (1982) *Technology and Competition in the Brazilian Computer Industry* (London: Frances Pinter).

Bator, F. M. (1958) 'The Anatomy of Market Failure', *Quarterly Journal of Economics*, pp. 351–79.

Becker, R. H. (1980) 'Project Selection Checklists for Research, Product Development and Process Development', *Research Management*, 23, p. 5.

Beed, C. (1966) 'The Separation of Ownership from Control', *Journal of Economic Studies*, pp. 29–46.

Beer, J. J. (1959) *The Emergence of the German Dye Industry* (Illinois University Press).

Benn, A. W. (1971) 'Technical Power and People: the Impact of Technology on the Structure of Government', *Bulletin of Atomic Scientists*, (December).

Berle, A. and G. Means (1932) *The Modern Corporation and Private Property* (New York: Macmillan).

Bertalanffy, L. von (1973) *General System Theory* (Harmondsworth: Penguin).

BIFU (1980) *Report of the Microelectronics Committee*, Banking, Insurance and Finance Union.

Binswanger, H., V. Ruttan, *et al.* (1978) *Induced Innovation* (Baltimore: Johns Hopkins University Press).

Bitondo, D. and A. Frohman (1981) 'Linking Technological and Business Planning', *Research Management*, 24 p. 6.

Blackburn, P., R. Coombs and K. Green (1985) *Technology, Economic Growth and the Labour Process* (London: Macmillan).

Bosworth, D. (1979) 'Recent Trends in R & D in the UK', *Research Policy*.

Bowen, H. (1980) 'US Comparative Advantage: A Review of the Theoretical and Empirical Advantage Literature, US Department of Labour (July) Mimeo.

Briggs, Asa (1954) *Victorian People* (Odhams Press).

Briggs, Asa (1961) 'Cholera and Society in the Nineteenth Century', *Past and Present*, No. 19 p. 76.

BSSRS (British Society for Social Responsibility in Science) (1983) *Hazards Bulletin*, No. 35 (May) p. 3.

Burns, T. and G. Stalker (1961) *The Management of Innovation* (London: Tavistock Publications).

Bush, V. (1947) *Endless Horizons* (Washington, DC) as quoted in E. Layton, 'Technology as Knowledge', *Technology and Culture*, 15 (1974), pp. 31–41.

Byalt I.C.R., and A.V. Cohen (1969) *An Attempt to Quantify the Economic Benefits of Scientific Research* (London: HMSO).

Carter, C. F. and B. R. Williams (1957) *Industrial and Technical Progress* (Oxford University Press).

Caves, R. *et al.* (1980) *Competition in the Open Economy* (Harvard University Press)

Caves, R. E. (1982) *Multinational Enterprise and Economic Analysis*, (Cambridge University Press).

Central Policy Review Staff (CPRS) (1971) *A Framework for Government Research and Development*, Cmnd 4814 (The Rothschild Report) (London: HMSO).

Central Policy Review Staff (1975) *The Future of the British Motor Industry* (London: HMSO).

Central Policy Review Staff (1978) *Social and Employment Implications of Microelectronics* (NEDC).

Chandler, A. D. (1962) *Strategy and Structure* (MIT Press).

Chandler, A. D. and H. Daems (eds) (1980) *Managerial Hierarchies: Comparative Perspectives on the Rise of the Modern Industrial Enterprise* (Harvard University Press).

Channon, D. F. (1973) *The Strategy and Structure of British Enterprise*, (London: Macmillan Press).

Cliffe, W. H. (1963) 'A Historical Approach to the Dyestuffs Industry', *J. Soc. Dyers and Colourists*, 79, p. 353.

Collingridge, D. (1980) *The Social Control of Technology* (London: Frances Pinter).

Colman, D. and F. Nixson (1978) *Economics of change in Less Developed Countries* (Oxford: Philip Allan).

Committee of Inquiry on Industrial Democracy (1977) *Report*, (The Bullock Report), Cmnd 6706, (London: HMSO).

Committee of Inquiry into the Engineering Profession (1980) *Engineering*

Our Future (The Finniston Report), Cmnd 7794 (London: HMSO).
Committee of Inquiry into Human Fertilisation and Embryology (1984) *Report* (Warnock Report), Cmnd 9414, (London: HMSO).
Connor, S. (1984) 'New Guidelines on Safety at Work', *New Scientist*, 16 August.
Cook, J. and C. Kaufman (1982) *Portrait of a Poison* (London: Pluto Press).
Cooley, M. (1972) *Computer Aided Design* (AUEW [TASS]).
Coombs, R. (1984a) Long-term Trends in Automation' in P. Marstrand (ed.) *New Technology and the Future of Work and Skills* (London: Frances Pinter).
Coombs, R. (1984b) 'Long-waves and Labour-Process Change', *Review*, 7, p. 4.
Coombs, R. (1985) 'Automation, Management Strategies and Labour-Process Change' in D. Knights (ed.) *Job Redesign: Critical Perspectives on the Labour Process* (Farnborough: Gower Press).
Cornwall, J. (1977) *Modern Capitalism: Its Growth and Transformation* (Oxford: Martin Robertson).
Council for Science and Society (CSS) (1977) *The Acceptability of Risks* (London: Barry Rose).
Cowling, K. (1982) *Monopoly Capitalism* (London: Macmillan Press).
Crosland, A. (1952) 'The Transition from Capitalism' in R. Crossman (ed.) *New Fabian Essays* (London: Turnstile Press).
Cyert, R. M. and J. G. March (1963) *A Behavioural Theory of the Firm* (Englewood Cliffs NJ: Prentice-Hall).
Dalton, A. (1979) *Asbestos Killer Dust* (London: BSSRS Publications).
Dalyell, T. (1985) *New Scientist*, 8 August.
Darlington, R. (1979) *The Modernisation of Telecommunications*, (Post Office Engineering Union [now National Communications Union]).
David, P. (1975) *Technical Choice, Innovation and Economic Growth*, (Cambridge University Press).
Davies, S. (1979) *The Diffusion of Process Innovations* (Cambridge University Press).
Denison, E. (1962) 'United States Economic Growth', *Journal of Business*, vol. 35, pp. 109–21.
Department of Industry (1975a) *The Regeneration of British Industry*, Cmnd. 5710 (London: HMSO).
Department of Industry (1975b) *British Leyland – the Next Decade*, (London: HMSO).
Doll, R. (1955) 'Mortality from Lung Cancer in Asbestos Workers', *British Journal of Industrial Medicine*, 12 p. 817.
Domar, D. (1946) 'Capital Expansion, Rate of Growth and Employment', *Econometrica*, vol. 14.
Dosi, G. (1981) *Structural adjustment and public policy under conditions of rapid technical change: the semi-conductor industry in Western Europe* (Brighton: University of Sussex European Research Centre).
Dosi, G. (1982) 'Technological Paradigms and Technological Trajectories: a Suggested Interpretation of the Determinants and Directions of Technical Change', *Research Policy*, 11.

Dosi, G. (1984b) 'Technology and Conditions of Macroeconomic Development' in C. Freeman (ed.), *Design, Innovation and Long Cycle in Economic Development* (London: Design Research Publications).

Dosi, G. (1984a) *Technological Change and Industrial Transformation* (London: Macmillan).

Dunning, J. H. (1979) 'Explaining Changing Patterns of International Production: in Defence of the Eclectic Theory' *Oxford Bulletin of Economics and Statistics*, 41, pp. 269–295.

Dunning, J. M. (1980) 'Toward an Eclectic Theory of International Production: Some Empirical Tests', *Journal of International Business Studies*, 11 (1980), pp. 9–31.

Elliott, D. and R. Elliott (1976) *The Control of Technology* (London: Wykeham Publications, 1976).

Finer, S. E. (1952) *The Life and Times of Sir Edwin Chadwick* (London: Methuen).

Ford, G. (1985) *New Scientist*, 25 July.

Freeman, C. (1979) 'The Determinants of Innovation', *Futures*, II, p. 206.

Freeman, C. (1982) *The Economics of Industrial Innovation* (London: Frances Pinter).

Freeman, C., J. Clark, and L. Soete (1982) *Unemployment and Technical Innovation* (London: Frances Pinter).

Freeman, C. (1984) 'Keynes or Kondratiev? How can we get back to full employment?' in P. Marstrand (ed.) *New Technology and the Future of Work and Skills* (London: Frances Pinter).

Galbraith, J. K. (1967) *The New Industrial State* (Harmondsworth: Penguin Books).

GMAG (1978) (Genetic Manipulation Advisory Group), *First Report*, Cmnd 7215 (London: HMSO).

Gibbons, M. and R. Johnston (1974) 'The roles of science in technological innovation', *Research Policy*, 3, pp. 220–242.

Gibbons, M., R. Coombs, P. Saviotti, P. C. Stubbs, (1982) 'Innovation and Technical Change: a Case Study of the UK Tractor Industry, *Research Policy*, 11 pp. 289–310.

GMBATU (1980) *New Technology* (General, Municipal, Boilermakers and Allied Trades Union).

Grabowski, H. G. (1976) *Drug Regulation and Innovation* (American Enterprise Institute).

Grasham, K. (1985) Letter to M. Thatcher quoted by *New Scientist*, 25 July.

Green, K. (1982) 'Health and Safety at Work and the Radical Science Movement' in D. Elliott, K. Green and F. Steward (eds) *Trade Unions, Technology and the Environment* (Milton Keynes: Open University Press).

Green K. and C. Morphet (1977) *Research and Technology as Economic Activities* (London: Butterworth).

Griliches, Z. (1957) 'Hybrid Corn: an Exploration in the Economics of Technical Change', *Econometrica*.

Gordon, D., R. Edwards and M. Reich (1982) *Segmented Work, Divided*

Workers, The Historical Transformation of Labour in the United States (Cambridge University Press).

Gough, I. (1979) *The Political Economy of the Welfare State* (London: Macmillan Press).

The *Guardian*, 18 October 1985. Various articles on the Lawlords' ruling on the 'Gillick Case'.

Gummett, P. (1980) *Scientists in Whitehall* (Manchester University Press).

Gummett, P. (1983) 'Customers and Contractors: Recent British Experience in the Commissioning of Research by Government' in *Research Ethics*, K. Berg and K. E. Tranoy (eds) (New York: Alan R. Liss).

Habbakuk, H. (1962) *American and British Technology in the XIXth Century* (Cambridge University Press).

Haber, L. F. (1958) *The Chemical Industry During the Nineteenth Century* (Oxford University Press).

Hannah, L. (ed.) (1976) *Management Strategy and Business Development: An Historical and Comparative Study* (London: Macmillan Press).

Hardie, D. W. F. (1950) *A History of the Chemical Industry in Widnes* (Imperial Chemical Industries).

Harrod, R. F. (1949) 'An Essay in Dynamic Theory', *Economic Journal* vol. 49, no. 1.

Hart, A. (1966) 'A chart for evaluating R & D projects', *Operational Research Quarterly*, 17, p. 4.

Hay, D. (1983) 'Management and Economic Performance' in M. Earl (ed.) *Perspectives on Management* (Oxford University Press).

Hayvaert, C. H. (1973) *Innovation Research and Product Policy: Clinical Research in 12 Belgian Industrial Enterprises* (Belgium: Catholic University of Louvain).

Health and Safety Executive (1984) Guidance Note EH 10/84, *Asbestos - Control Limits*.

Heertje, A. (1977) *Economics and Technical Change* (London: Weidenfeld & Nicolson).

Hetman, F. (1973) *Society and the Assessment of Technology* (Paris: OECD).

Hicks, J. R. (1932) *The Theory of Wages* (London: Macmillan).

Hildrew, P. (1982) 'Shop Stewards Fight "Typically British" Job Cuts', The *Guardian*, 2 March.

Hodgson, G. (1984) *The Democratic Economy* (Harmondsworth: Penguin).

Hollomon, J. H. (1979) 'Policies and Programs of Governments Directed Towards Industrial Innovation' in C. T. Hill and J. M. Utterback (eds) *Technological Innovation for a Dynamic Economy* (Oxford: Pergamon).

Horwitz, P. (1979) 'Directing Government Funding of R & D: Intended and Unintended Effects on Industrial Innovation', in C. T. Hill and J. M. Utterback (eds) *Technological Innovation for a Dynamic Economy* (Oxford: Pergamon).

Huws, U. (1982) *Your Job in the Eighties* (London: Pluto Press).

284 *Bibliography*

Irvine, J. and B. Martin (1984) 'What Direction for Basic Scientific Research?' in M. Gibbons, P. Gummett and B. M. Udgaonkar, *Science and Technology Policy in the 1980s and Beyond* (London: Longman).

Irwin, A. (1985) *Risk and Control of Technology* (Manchester University Press).

Jevons, F. R. (1973) *Science Observed* (London: George Allen & Unwin).

Jewkes, J. (1948) *Ordeal by Planning* (London: Macmillan).

Jewkes, J. (1958) D. Sawers and R. Stillerman, *The Sources of Invention* (London: Macmillan).

Johnston, R. and P. Gummett (eds) (1979) *Directing Technology* (London: Croom Helm).

Jones, T. T. and T. A. J. Cockerill (1984) 'Public Enterprise Management' in J. F. Pickering and T. A. J. Cockerill, *Economic Management of the Firm* (Oxford: Philip Allan).

Joseph, K. (1970) Speech 7 March reported in *The Sunday Times*, 8 March.

Joyce, C. (1985) 'Science under Reagan: the First Four Years', *New Scientist*, 24 January.

Kaldor, N. (1966) *Causes of the Slow Rate of Economic Growth of the United Kingdom* (Cambridge University Press).

Kaldor, M. (1980) 'Technical Change in the Defence Industry', in K. Pavitt (ed.), *Technical Change and British Economic Competitiveness* (London: Macmillan).

Kamien, M. I. and N. L. Schwartz (1982) *Market Structure and Innovation* (Cambridge University Press).

Kaufman, G. (1984) Personal Communication.

Kay, N. (1979) *The Innovating Firm* (London: Macmillan).

Kay, N. (1982) *The Emergent Firm* (London: Macmillan).

Kennedy, C. (1964) 'Induced Bias in Innovation and the Theory of Distribution', *Economic Journal*, 74, pp. 541–7.

Kennedy, D. M. (1970) *Birth Control in America* (Yale University Press).

Keynes, J. M. (1936) *General Theory of Employment Interest and Money* (New York: Harcourt, Brace).

Kindleberger, C. P. (1965) *Economic Development*, (New York: McGraw-Hill). (Also 3rd edn, B. M. Herrick.)

Klein, B. (1977) *Dynamic Economics* (Harvard University Press).

Kleinknecht, A. (1984) 'Innovation Patterns in Crisis and Prosperity: Schumpeter's Long Cycle Reconsidered', doctoral thesis (Free University of Amsterdam).

Knight, F. H. (1965) *Risk, Uncertainty and Profit* (New York: Harper & Row).

Kondratiev, N. (1978) 'The Major Economic Cycles', *Lloyds Bank Review*, 129 (reprinted from *Review of Economic Statistics*, 18).

Kuznets, S. (1930) *Secular Movements in Production and Prices* (Boston: Houghton Mifflin).

Labour Party (1963) *Report of Annual Conference*, (London: Labour Party).

Lall, S. (1980) *The Multinational Corporation* (London: Macmillan).

Langrish, J., M. Gibbons, W. Evans, and F. Jevons (1972) *Wealth from Knowledge* (London: Macmillan).

Leamer, E. (1980) 'The Leontieff Paradox reconsidered', *The Journal of Political Economy*, vol. 88 pp. 495–513.

Legraw, D. J. (1984) 'Diversification Strategy and Performance', *J. Industrial Economies*, 33, p. 2.

Leontieff, W. (1953) 'Domestic Production and Foreign Trade: the American Position Re-examined', *Proceedings of the American Philosophical Society*, vol. 97.

Lesser, F. (1983) 'Drugs Monitor Needs Sharper Teeth', *New Scientist*, 17 March.

Levinstein, I. (1883) *Journal of the Society of Chemical Industries*, 2, 213.

Lilley, S. (1973) 'Technological Progress and the Industrial Revolution, 1700–1914', in C. Cipolla (ed.) *The Industrial Revolution* (London: Fontana).

Littler, D. and R. Sweeting, (1983) 'New Business Development in Mature Firms', *Omega*, 11 (6), p. 537.

Lloyd, I. (1985) *New Scientist*, 4 July.

Loveday, D. E. (1984) 'Factors affecting the management of interdisciplinary research in the pharmaceutical industry', *R & D Management*, 14, (2).

MacDonald, S., D. Lamberton, T. D. Mandeville (1984) *The Trouble with Technology* (London: Frances Pinter).

McGinty, L. (1979) *New Scientist*, 9 August, p. 433.

MacKenzie, D. (1983 and 1984) *New Scientist*, 8 September 1983, and 6 September 1984.

McLeod, R. M. (1965) *Victorian Studies*, IX, 85.

Machlup, F. (1967) 'Theories of the Firm: Marginalist, Behavioural and Managerial', *American Economic Review* 57.

Maddison, A. (1964) *Economic Growth in the West* (London: George Allen & Unwin).

Maddison, A. (1982) *Phases of Capitalist Development* (Oxford University Press).

Magee, S. P. (1977) 'Information and Multinational Corporations: an Appropriability Theory of Direct Foreign Investment', in J. Bhagwati (ed.), *The New International Economic Order* (Cambridge, MA: MIT Press).

Malthus, T. (1920) *First Essay on Population* (London: Royal Economic Society). 1st pub. 1798.

Malthus, T. R. (1976) *An Essay on the Principle of Population* (Reprinted, Harmondsworth: Penguin).

Mandel, E. (1969) *An Introduction to Marxist Economic Theory* (New York: Pathfinder Press).

Mandel, E. (1975) *Late Capitalism* (London: New Left Books).

Mansfield, E. (1963) 'Intra-firm Rates of Diffusion of an Innovation', *Review of Economics and Statistics*, 30.

Mansfield, E. (1969) *Industrial Research and Technological Innovation: an econometric analysis* (London: Longman).

Mansfield, E., J. Rapoport, A. Romeo, S. Wagner and G. Beardsley (1977) 'Social and Private Rates of Return from Industrial Innovations', *Quarterly Journal of Economics*, May.

Mansfield, E., A. Romeo and S. Wagner (1979) 'Foreign Trade and US Research and Development, *Review of Economics and Statistics*, 61, pp. 49–52.

Marris, R. (1966) *The Economic Theory of 'Managerial' Capitalism* (London: Macmillan).

Martin, B., and J. Irvine (1984) *Foresight in Science* (London: Frances Pinter).

Martin, B., J. Irvine and R. Turner (1984) 'The Writing on the Wall for British Science', *New Scientist*, 8 November.

Marx, K. (1961) *Capital* (Moscow: Foreign Language Publishing House) 1st pub. 1867.

Mathias, P. (1984) 'The Machine, icon of economic growth' in S. MacDonald, D. McL. Lamberton, T. D. Mandeville (eds) *The Trouble with Technology* (London: Frances Pinter).

Mensch, G. (1979) *The Technological Stalemate* (New York: Ballinger).

Merewether, E. R. A. and C. W. Price (1934) *Effects of Asbestos Dust on the Lungs and Dust Suppression in the Asbestos Industry* (London: HMSO.

MERG, (1982) *Leyland Vehicles: The Workers' Alternative* (Leyland Joint Works Committee).

Merrifield, D. B. (1981) 'Selecting Projects for Commercial Success', *Research Management*, 24, p. 6.

Merton, R. K. (1942) 'Science and Technology in a Democratic Order', *Journal of Legal and Political Sociology*, 1.

Metcalfe, J. S. (1970) 'The Diffusion of Innovations in the Lancashire Textile Industry', *Manchester School of Economics and Social Studies*, 2, pp. 145–62.

Metcalfe, J. S. (1981) 'Impulse and Diffusion in the Study of Technical Change', *Futures*, 13, pp. 347–59.

Miles, R. and C. Snow (1978) *Organisation Strategy, Structure and Process*, (New York: McGraw-Hill).

Momigliano, F. and G. Dosi, (1983) *Tecnologia and Organizzazione Industriale Internazionale* (Bologna: Il Mulino).

Moss, S. (1981) *An Economic Theory of Business Strategy* (Oxford: Martin Robertson).

Mowery, D. and N. Rosenberg (1979) 'The Influence of Market Demand upon Innovation: a Critical Review of Some Recent Empirical Studies', *Research Policy*, 8, pp. 102–53.

Muncaster, J. M. (1981) 'Picking New Product Opportunities', *Research Management*, 24, p. 4.

Myers, S. and D. G. Marquis (1969) *Successful Industrial Innovation: a Study of Factors Underlying Innovation in Selected Firms* (Washington: National Science Foundation, NSF 69–17).

Nabseth, L. and G. Ray (eds) (1974) *The Diffusion of New Industrial Processes* (Cambridge University Press).

Nelkin, D. (1977) *The Politics of Participation* (Beverly Hills: Sage Publications).

Nelson, R. (1981) 'Research on Productivity Growth and Productivity Differences: dead ends and new departures', *Journal of Economic Literature*, vol. 19 (1981), pp. 1029–1064.

Nelson, R. and S. Winter (1974) 'Neoclassical and Evolutionary Theories of Growth: Critique and Prospectus', *Economic Journal*, pp. 886–905.

Nelson, R. and S. Winter (1977) 'In search of useful theory of innovation', *Research Policy*, 6, pp. 36–76.

New Scientist (1985a), 1 August, 'This Week'.

New Scientist (1985b), 13 June, 'This Week'.

New Scientist (1985c), 25 July, 'This Week: Ministers call for survey of beach sewage'.

Norris, K. and J. Vaizey (1973) *The Economics of Research and Technology* (London: George Allen & Unwin).

OECD (1971) *Science, Growth and Society* (Paris).

OECD (1980) *Technical Change and Economic Policy* (Paris).

OECD (1982) *Workshop on Patent and Innovation Statistics, Summary of Contributions* (Paris).

OECD (1985) *Recent Results: Selected Science and Technology Indicators, 1981–6*, October.

Parliamentary Commission for Administration, (1976) *Third Report*: Session 1975–6, H-C259 (London: HMSO) pp. 189–211.

Pasinetti, L. L. (1981) *Structural Change and Economic Growth* (Cambridge University Press).

Pavitt, K. (1980) 'Introduction' in K. Pavitt (ed.), *Technical Innovation and British Economic Performance* (London: Macmillan).

Pavitt, K. (ed.) (1980) *Technical Innovation and British Economic Performance* (London: Macmillan).

Pavitt, K. L. R. and L. Soete, (1980) 'Innovative Activities and Export Shares: some comparisons', in K. Pavitt (ed.) *Technical Innovation and British Economic Performance* (London: Macmillan).

Pearce, F. (1984) 'The Great Drain Robbery', *New Scientist*, 15 March.

Peltzman, S. (1974) *Regulation of Pharmaceutical Innovation* (American Enterprise Institute).

Penrose, E. (1980) *The Theory of the Growth of the Firm* (Oxford: Blackwell).

Perez, C. (1983) 'Structural Change and the Assimilation of New Technologies in the Economic and Social System', *Futures*, October.

Petersen R. and S. E. Venstrup-Nielsen, (1982) *Mineraluld-Symptom Belasting og Arbejdsforhold ved isoleringsarbejde med Mineraluld*, Kobenhavn, Aarhus, Odense: FADLIS Forlag. We are grateful to the authors for discussing their results with us in English!

Pickering, J. F. and T. T. Jones (1984) 'The Firm and its Social Environment' in J. F. Pickering and T. A. J. Cockerill, *Economic Management of the Firm* (London: Philip Allan).

Polanyi, K. (1957) *The Great Transformation* (Boston, Mass: Beacon).

Poole, J. B. and K. Andrews (1972) *The Government of Science in Britain*, (London: Weidenfeld & Nicolson).

Posner, M. (1961) 'International trade and technical change', *Oxford Economic Papers*, vol. 13, pp. 323–41.

Posner, M. V. and A. Steer (1979) 'Price Competitiveness and Export Performance' in F. Blackaby (ed.), *De-industrialisation* (London: Heinemann Educational Books).

Reppy, J. (1979) 'The Control of Technology through Regulation', in R. Johnston and P. Gummett (eds) *Directing Technology* (London: Croom Helm).

Reuben, B. R. and M. Burstall (1973) *The Chemical Economy* (London: Longman).

Ricardo, D. (1813) *The Principles of Political Economy and Taxation* (first published 1813).

Rogers, E. (1962) *Diffusion of Innovations* (New York: Collier-Macmillan).

Rose, H. and S. Rose (1970) *Science and Society* (Harmondsworth: Penguin).

Rosenberg, N. (1969) 'The Direction of Technical Change: Mechanisms and Focusing Devices', *Economic Development and Cultural Change*, 18.

Rosenberg, N. (1974) 'Science Invention and Economic Growth', *Economic Journal*, 84.

Rosenberg, N. (1976) 'The Direction of Technological Change, Inducement Mechanisms and Focusing Devices', in *Perspectives on Technology* (Cambridge University Press).

Rosenberg, N. (1982) *Inside the Black Box: Technology and Economics* (Cambridge University Press).

Rosenberg, N. and C. Frischtak (1984) 'Technological Innovation and Long Waves', *Cambridge Journal of Economics*, 8 (1).

Rosseger, G. E. (1980) *The Economics of Production and Innovation* (London: Pergamon).

Rothschild, (1982) *An Enquiry into the Social Science Research Council*, Cmnd 8554 (London: HMSO).

Rothwell, R. (1976) *Innovation in Textile Machinery: Some significant factors in success and failure* (Brighton: Science Policy Research Unit, Occasional Paper No. 2).

Rothwell, R. (1977) 'The characteristics of successful innovators and technically progressive firms (with some comments on innovation research), *R & D Management*, 7 (3) pp. 191–206.

Rothwell, R. and V. Walsh (1979) 'Regulation and Innovation in the Chemical Industry' (OECD, mimeo).

Rothwell, R. and W. Zegveld, (1981) *Industrial Innovation and Public Policy* (London: Frances Pinter).

Roy, R. (1978) *Social Control of Technology* (Milton Keynes: Open University Press; new edition 1982).

Royal Commission on Environmental Pollution (RCEP) (1976) *Sixth*

Report: Nuclear Power and the Environment, Cmnd 6618 (London: HMSO).

Sahal, D. (1981) 'Alternative Conceptions of Technology', *Research Policy*, 10, pp. 2–24.

Salter, W. (1966) *Productivity and Technical Change*, 2nd edition (Cambridge University Press).

Saren, M. A. (1984) 'Classification and Review of Models of the Intrafirm Innovation Process', *R & D Management*, 14, 1.

Saviotti, P., R. W. Coombs, M. Gibbons and P. C. Stubbs, (1980) *Technology and Competitiveness in the Tractor Industry*, A Report for the Department of Industry (Manchester).

P. P. Saviotti and A. Bowman (1984) 'Indicators of Output of Technology' in M. Gibbons, P. Gummett and B. M. Udgaonkar, (Eds) *Science and Technology Policy in the 1980s and Beyond*, (London: Longman).

Saviotti, P. P. and J. S. Metcalfe (1984) 'A Theoretical Approach to the Construction of Technological Output Indicators', *Research Policy*, 13, pp. 141–51.

Sbragia, R. (1984) 'Clarity of Manager Roles and Performance of R & D Multi-Disciplinary Projects in Matrix Structures', *R & D Management*, 14 (2).

Scherer, F. M. (1980) *Industrial Market Structure and Economic Performance*, 2nd edition (Chicago: Rand McNally).

Schmidt, A. (1974) Testimony to US Senate Subcommittee on Health, of the Committee on Labor and Public Welfare. *Hearings on Legislation Amending the Public Health Service Act and the Federal Food, Drugs and Cosmetic Act*, 93rd Congress (US Government Printing Office), pp. 3077–3103.

Schmookler, J. (1966) *Invention and Economic Growth* (Harvard University Press).

Schumpeter, J. A. (1934) *The Theory of Economic Development* (Cambridge, Mass.: Harvard University Press).

Schumpeter, J. A. (1939) *Business Cycles* (New York: McGraw-Hill); second edition 1964.

Schumpeter, J. A. (1942) *Capitalism, Socialism and Democracy* (London: Allen & Unwin, 1976; first edition New York: Harper & Row, 1942).

Schumpeter, J. A. (1943) *Capitalism, Socialism and Democracy* (New York: Harper & Row).

Science Policy Research Unit (SPRU) (1972) *Success and Failure in Industrial Innovation* (London: Centre for the Study of Industrial Innovation).

Seal, V. (1968) 'Industrial Innovation: Case Studies of Seven Recipients of the Queens Award to Industry', MSc thesis, Manchester University.

Sehock, G. (1974) *Innovation Processes in Dutch Industry*, T.N.O., Industrial Research Organisation, Apeldoorn, The Netherlands.

Sherwin, C. W. and R. S. Isenson, (1967) 'Project Hindsight', *Science*, 156 pp. 1571–7.

Shonfield, A. (1981) 'Innovation: Does Government Have a Role?' in G.

Carter (ed.) *Industrial Policy and Innovation* (London: Heinemann).

Shore, P. (1978) Statement 7.3.77 quoted in G. Boyle, *Nuclear Power – the Windscale Controversy* (Milton Keynes: Open University).

Smith, Adam (1776) *The Wealth of Nations* (Harmondsworth: Penguin Books, 1970).

Social Audit (1974) *The Alkali Inspectorate* (London: Social Audit).

Soete, L. (1979) 'Firm Size and Inventive Activity: the evidence reconsidered', *European Economic Review*.

Soete, L. (1981) 'A general test of technology gap trade theory' in *Weltwirtschaftliches Archiv*, vol. 117, pp. 638–660.

Soete, L. (1982) *Innovation and International Trade* (Brighton: SPRU).

Soete, L. and G. Dosi (1983) *Technology and Employment in the Electronics Industry* (London: Frances Pinter).

Sohn-Rethel, A. (1978) *Economy and Class Structure of German Fascism* (London: CSE Books).

Solow, R. (1957) 'Technical change and the aggregate production function', *Review of Economics and Statistics*, vol. 39, pp. 312–20.

Stern, R. and K. Maskus (1981) 'Determinants of the Structure of US Foreign Trade, 1958–1976', *Journal of International Economics*, vol. 11 p. 207–224.

Steward, F. (1978) 'Public Policy and Innovation in the Drug Industry,' in Sir Douglas Black and G. P. Thomas (eds) *Providing for the Health Services*, BAAS (X), 1977 (London: Croom Helm).

Steward, F. and G. Wibberley, (1980) 'Drug Innovation – Why is it Slowing Down?', *Nature*, 284, pp. 118–120.

Stubbs, P. (1979) 'Technology Policy and the Motor Industry', in R. Johnston and P. Gummett (eds) *Directing Technology* (London: Croom Helm).

Stoneman, P. (1976) *Technological Diffusion and the Computer Revolution* (Cambridge University Press).

Stoneman, P. (1983) *The Economic Analysis of Technological Change* (Oxford University Press).

Stoneman, P. (1984) 'Theoretical approaches to the analysis of the diffusion of new technology' in S. MacDonald *et al.*, *The Trouble with Technology* (London: Frances Pinter).

Stout, D. K. (1977) *International Competitiveness, Non-price Factors and Export Performance* (London: NEDO).

Sweden, (1981) National Board of Occupational Safety and Health, *Criteria Document for Swedish Occupational Standards: Asbestos and Inorganic Fibres* (in English).

Teece, D. J. (1976) *The Multinational Corporation and the Resource Cost of International Technology Transfer* (Cambridge, MA: Ballinger).

Teece, D. J. (1977) 'Technology Transfer by Multinational Firms: the Resource Cost of Transferring Technological Know-how', *Economic Journal*, 87, pp. 242–61.

Teece, D. J. (1981) 'The Market for Know-how and Efficient International Transfer of Technology', *Annual of the American Academy of Political and Social Science* 458 pp. 81–96.

Teeling-Smith, G. (1969) 'Relationships with Government' in G. Teeling-Smith (ed.) *Economics and Innovation in the Pharmaceutical Industry* (Office of Health Economics).

Thomas, H. (1970) *Econometric and Decisions Analysis: Studies in R & D in the Electronics Industry*, Ph D Thesis (University of Edinburgh).

The Times, 1 August 1858.

Tomlinson, J. (1982) *The Unequal Struggle* (London: Methuen).

TRACES (1968) - *Technology in Retrospect and Critical Events in Science* (Washington: National Science Foundation).

Twiss, B. (1980) *The Management of Technological Innovation* (London: Longman).

Urquhart, J. and K. Heilmann (1985) *Risk Watch* (New York: Facts on File).

Utterback, J. M. *et al.*, (1973) *The Process of Innovation in Five Industries in Europe and Japan* (Cambridge, Mass.: Centre for Policy Alternatives).

Utterback, J. M. and W. J. Abernathy, (1975) 'A Dynamic Model of Product and Process Innovation', *Omega*, 3, (6) pp. 639–656.

Vernon, R. (1966) 'International Investment and International Trade in the Product Cycle', *Quarterly Journal of Economics*, 80, pp. 190–207.

Wainwright, H. and D. Elliot, (1982) *The Lucas Plan* (London: Allison & Busby).

Walker, W. (1980) 'Britain's Economic Performance, 1850–1950', in K. Pavitt (ed.) *Technical Innovation and British Economic Performance* (London: Macmillan).

Walsh, V. (1980) 'Contraception, the Growth of a Technology' in L. Birke *et al.*, *Alice Through the Microscope* (London: Virago).

Walsh, V. (1984) 'Invention and innovation in the chemical industry: Demand-pull or discovery-push?', *Research Policy*, 13, pp. 211–234.

Walsh, V., J. Townsend, B. Achilladelis, C. Freeman (1979) *Trends in Innovation and Innovation in the Chemical Industry* (Brighton: University of Sussex).

Walsh, V. and R. Roy, (1983) *Plastics Products: Good Design Innovation and Business Success* (Milton Keynes: Open University Press).

Walsh, V., J. Moulton-Abbott and P. Senker, (1980) *New Technology, The Post Office and the Union of Post Office Workers*, Union of Post Office Workers (now Union of Communication Workers).

Weiszacker, C. von (1966) 'Tentative Notes on a Two-Sector Model with Induced Technical Progress', *Review of Economic Studies*, xxxiii, pp. 245–51.

Wells, L. T. (1972) *The Product Lifecycle and International Trade*, (Boston: Harvard University, Graduate School of Business Administration, Division of Research Publications).

Werskey, G. (1978) *The Visible College* (London: Allen Lane).

Wheare, K. C. (1955) *Government by Committee* (Oxford University Press).

Whitley, R. (1984) 'Theories of firm behavior' (mimeo) (Manchester Business School).

Whitley, R. (1984b) *The Intellectual and Social Organisation of the Sciences* (Oxford University Press).

Williams, R. (1971) *Politics and Technology* (London: Macmillan).

Williams, R. (1973) *Technology and Society*, 8, p. 65.

Williams, R. (1983–4) 'British Technology Policy, *Government and Opposition*, Winter pp. 30–51.

Williams, R. R. Roy and V. Walsh(1982), *Government and Technology* (Open University Press, 2nd edition).

Williams, S. (1971) 'The Responsibility of Science', *The Times*, 27 February.

Williamson, O. E. (1962) *The Economics of Discretionary Behaviour: Management Objectives in a Theory of the Firm* (Englewood Cliffs, NJ: Prentice-Hall 1964).

Williamson, O. E. (1967) 'Hierarchical Control and Optimum Firm Size', *Journal of Political Economy*, 75.

Williamson, O. E. (1975) *Markets and Hierarchies: Analysis and Antitrust Implications* (New York: Free Press).

Williamson, O. E. (1981b) 'The Economics of Organisation: the Transaction Costs Approach', *American Journal of Sociology*, no. 87 (November), pp. 548–77.

Williamson, O. E. (1981a) 'The Modern Corporation', *Journal of Economic Literature* (December).

Willott, W. B. (1981) 'The NEB Involvement in Electronics and Information Technology' in C. Carter (ed.) *Industrial Policy and Innovation* (London: Heinemann).

Wilson, D. (1983) *The Lead Scandal* (London: Heinemann).

Wragg, R. and T. Robertson (1978) *Post-war Trends in Employment, Productivity, Output, Labour Costs and prices by Industry in the UK*, Research Paper No. 3 (London: Department of Employment).

Index